St Helena

St Helena
An Island Biography

Arthur MacGregor

THE BOYDELL PRESS

© Arthur MacGregor 2024

All Rights Reserved. Except as permitted under current legislation no part of this work may be photocopied, stored in a retrieval system, published, performed in public, adapted, broadcast, transmitted, recorded or reproduced in any form or by any means, without the prior permission of the copyright owner

The right of Arthur MacGregor to be identified as the author of this work has been asserted in accordance with sections 77 and 78 of the Copyright, Designs and Patents Act 1988

First published 2024
The Boydell Press, Woodbridge

ISBN 978 1 83765 088 0

The Boydell Press is an imprint of Boydell & Brewer Ltd
PO Box 9, Woodbridge, Suffolk IP12 3DF, UK
and of Boydell & Brewer Inc.
668 Mt Hope Avenue, Rochester, NY 14620–2731, USA
website: www.boydellandbrewer.com

A CIP catalogue record for this book is available
from the British Library

The publisher has no responsibility for the continued existence or accuracy of URLs for external or third-party internet websites referred to in this book, and does not guarantee that any content on such websites is, or will remain, accurate or appropriate

This publication is printed on acid-free paper

Contents

List of Illustrations	vii
Photo credits	xi
Preface	xiii
Acknowledgements	xv

1	Genesis	1
	The island and its setting	1
	The loss of Eden	6
2	The breach: Europe and St Helena collide	8
	The Portuguese, Dutch, and the British	9
3	Population and environment: early impacts	17
	Animal and botanical resources	18
4	Population and environment: asserting control	30
	Populating the island	30
	Slaves and near-slaves	34
	Feeding the population	44
	Controlling the natural economy	56
5	'The citadel of the South Atlantic'	65
	Securing the island	66
	Defences and defenders	68
	Signals and telegraph	83
6	Scientists in transit: St Helena as a site for scientific investigation	107
	A constellation of astronomers	107
	St Helena and the magnetic crusade	119
	Wind, weather and tide at St Helena	124
	A succession of naturalists	126
7	Napoleon on St Helena	150

Contents

8	Later detainees, 1800s and 1900s	166
	Policing the slave trade	167
	Zulu chieftains in the mid-Atlantic	172
	Boer War prisoners	174
	Sayyid Khalid bin Barghash Al-Busa'ldi	182
	An unfulfilled Irish interlude	185
	The Bahraini Three	186
	Asylum revisited	188
9	A place in the modern world	190
	The coming of the submarine telegraph	190
	Radio and television	192
	Maritime oblivion again	193
	Birth of an airport	198
	Biosecurity and regeneration in the new international age	202
	And next: Utopia in the twenty-first century	204

Appendix: Governors of St Helena — 207

Bibliography — 209
Index — 219

Illustrations

Figures

1	Theodore de Bry, *The Island of St Helena* (1601)	2
2	Frontispiece from Francis Godwin's *The Man in the Moone* (1638)	7
3	Theodore de Bry, *The Island of St Helena* (1601) (detail)	11
4	Portuguese mariners shooting birds and goats; unattributed Portuguese map (1600s) (detail)	23
5	Peter Mundy's encounter with a 'sealionness' (1656)	28
6	Good conduct medal awarded to an enslaved person (1824)	38
7	Enslaved labourers tilling the ground by hand at Brooke Hill Farm (1809)	39
8	Poster for the sale and letting of slaves on St Helena (18 May 1829)	41
9	Windmill erected at Longwood (c.1858)	46
10	Boer prisoners of war employed in the whale oil industry (1900–02)	55
11	Samuel Thornton, *A Prospect of James Fort* (1702–07)	69
12	Map of the fortifications and batteries on St Helena (2006)	71
13	James Fort, plan views of four principal building phases (2013)	75
14	Badges of St Helena Local Militia and St Helena Rifles	78
15	Flamsteed's sextant, on which Halley's instrument, brought to St Helena, was based (1725)	110
16	Halley's observatory (1677–78)	111
17	Ladder Hill observatory (1835)	116
18	Time-ball of the observatory (1835)	118
19	Ground plan and elevation of the Longwood observatory (c.1847)	123
20	(a) William Burchell, (b) William Roxburgh, (c) Charles Darwin, (d) Joseph Dalton Hooker	134
21	Botanic garden, Jamestown (1810)	138
22	General Sir Hudson Lowe	156

Illustrations

23	The body of Napoleon Bonaparte laid out after death (1821)	161
24	Constructional details of Napoleon's tomb	163
25	Rupert's Valley, site of the Liberated Africans' camp (c.1858)	169
26	Excavation of the Liberated Africans' camp, Rupert's Valley (2007–08)	171
27	Dinuzulu takes his leave of the governor of St Helena (1897)	173
28	Boer prisoners of war arrive in Jamestown (1900–01)	176
29	The workshop of H. T. Siglé in the Deadwood Boer prisoner-of-war camp (1900–02)	178
30	Boer moneybox, signed 'K. Gey, St Helena. 1901'	179
31	Boer War cemetery at Knollcombes	182
32	Sayyid Khalid Barghash Al-Busa'ldi, Sultan of Zanzibar (1874–1927)	184
33	Cable-laying ship *Anglia* (1999)	191
34	RMS *St Helena* executing replenishment at sea (1983–4)	196
35	Charles Darwin's sketch of the updraught (1836)	201

Plates

1	George Hutchins Bellasis, *St Helena, taken from the Sea* (1815)	86
2	Plan of St Helena by James Imray & Son (1803–70)	87
3	Chart by James Rennell, including Atlantic currents and trade winds (1799)	88
4	Historic shipping routes, 1750–1800, compiled by Professor James Cheshire	88
5	Vestige of the Great Forest cover at Diana's Peak	89
6	Eroded coastal landscape with storm water courses	89
7	Porcelain salvaged from the wreck of the *Witte Leeuw*	90
8	Proclamation claiming St Helena for the Dutch Republic (15 April 1633)	91
9	The St Helena Wirebird (*Charadrius sanctaehelenae*) (1875)	92
10	Plantation House, the governor's country residence (1822)	92
11	New Zealand flax in present-day clearance campaign	93
12	Share certificate of St Helena Whale Fishery Company (23 December 1837)	93

Illustrations

13	Knoll Hill fort, completed in 1874	94
14	The Ladder Hill battery, mounted with an Elswick Mark VII coastal defence gun	94
15	George Hutchins Bellasis, the Jamestown approaches with multiple signal stations (1815)	95
16	F. R. Stack, *The rollers of 17 February 1846, taken from the Harbour Master's office*	96
17	Edmond Halley's star chart of the southern celestial hemisphere, engraved 1678	97
18	Edmond Halley's *New and Correct Chart shewing the Variations of the Compass in the Western and Southern Oceans* (1700)	97
19	William Burchell at the summit of Sugar Loaf (1807)	98
20	Illustrations from Charles Mellis's *St Helena* (1875): (a) variety of specimens, (b) *Dickinsonia arborescens*	98
21	The Briars, site of Napoleon's initial quarters on St Helena	100
22	Manual of signals indicating Napoleon's minute-by-minute situation (1815–16)	100
23	Longwood House following its expansion to accommodate its Napoleonic occupants	101
24	Napoleon's funeral: (a) The cortège sets out from Longwood; (b) it approaches the site of the tomb (1821)	102
25	Erich Mayer, *Camp life* in the Boer prisoner-of-war camp	104
26	Replica of a chessboard made by a Boer prisoner of war on St Helena	104
27	RMS *St Helena* (in service 1978–90), following her refit in 1978	105
28	RMS *St Helena* (in service 1990–2018), at anchor in James Bay	105
29	Boeing 757-200 landing at St Helena airport	106
30	Endemic Gumwood *Commidendrum robustum* growing in the Millennium Forest project (2023)	106

Full credit details are provided in the Photo Credits on p. xiii and captions to the images in the text. The author and publisher are grateful to all the institutions and individuals for permission to reproduce the materials in which they hold copyright. Every effort has been made to trace the copyright holders; apologies are offered for any omission, and the publisher will be pleased to add any necessary acknowledgement in subsequent editions.

Photo credits

Figures

Figs 1, 3: Eran Laor Cartographic Collection, The National Library of Israel (ct700); Fig. 2: © The British Library Board, C.58.c2, plate 15; Fig. 4: Biblioteca Nacional de Portugal (C.C. 343 P2); Fig. 5: Oxford, Bodleian Library, MS Rawlinson A 315, f.238, photo courtesy Malgosia Nowak-Kemp; Figs 6, 27: St Helena Museum; Figs 8, 9, 10, 16, 28, 31, 33, Plate 2: http://sainthelenaisland.info/; Fig. 11: Lionel Pincus and Princess Firyal Map Division, The New York Public Library; Figs 12, 13: from Ken Denholm: 12 *An Island Fortress*, 2006; 13 South Atlantic Fortress, http://sainthelenaisland.info/; Fig. 14: courtesy of Cultman Collectables; Figs 17, 18: *Nautical Magazine* 4 (1835), pl. opp. p. 658; Fig. 19: St Helena Archives; Fig. 20: (a) public domain, (b) Wikipedia, (c) Historic England/Bridgeman Images, (d) Wikimedia Commons; Fig. 22: from R. C. Seaton, *Sir Hudson Lowe and Napoleon* (1898); Fig. 23: Wellcome Collection, public domain; Fig. 26: courtesy of Andrew Pearson/the Rupert's Valley Archaeological Project; Figs 28, 29: Oorlogsmuseum van die Boererepublieke, Bloemfontein; Fig. 32: Public domain, https://flickr.com/photos/21022123@N04/48720358437/ (CC BY 2.0); Fig. 34: photograph by Robert A. Wilson, radio officer of the RMS, courtesy of John Bryant; Fig. 35: © Historic England Archive.

Plates

Plates 1, 15, 16: © National Maritime Museum, Greenwich; Plates 2, 11, 13, 14: http://sainthelenaisland.info/; Plate 3: Wikimedia Commons; Plate 4: courtesy of Professor James Cheshire; Plates 5, 6: St Helena Tourist Board; Plate 7: © Rijksmuseum, Amsterdam; Plate 8: © Nationaal Archief, The Hague (inv. no. 660); Plates 10, 21, 24(b): images courtesy of Professor Donal McCracken; Plates 11, 14: http://sainthelenaisland.info/; Plate 12: by kind permission of Spink and Son Ltd.; Plate 13: Marc Lavaud, http://sainthelenaisland.info/; Plates 17, 18, 23: courtesy of the British Library Board, Maps (17, Maps, 188.H.2(4); 18, Maps c.22.d.20, frontispiece; 23, MS Egerton 3717, f.178); Plate 20: courtesy of Linnean Society of London; Plate 22: University of Stirling Archives and Special Collections (Napoleonic Collection, MS 11); Plate 24(a): Wellcome Collection, public domain; Plate 25: Ditsong National Museum of Cultural History, Pretoria; Plate 26: St Helena Museum; Plate 27: photo Robert A. Wilson, courtesy of John Bryant; Plate 28: St Helena Government; Plate 29: What the Saints did Next, courtesy of Darrin and Sharon Henry; Plate 30: St Helena National Trust.

Preface

Admirable histories of St Helena have been compiled by a succession of extremely well-qualified authors over the past two centuries. In 1808, T. H. Brooke, sometime acting governor of the island, published his *History of the Island of St Helena, from its discovery by the Portuguese to the year 1806* (a second edition of 1824 brought it up to the previous year): Brooke's work has provided a starting point for every author since that time. Alexander Beatson, again governor for a time, followed on with his *Tracts relative to the Island of St Helena* (1816), still frequently consulted – especially for its many recommendations for agricultural reform – and in 1875 John Charles Mellis's *St Helena: a physical, historical, and topographical description of the island* added a naturalist's perspective to the empirical surveys of the island's progress. At the more recent end of the spectrum, Philip Gosse's *St Helena 1502–1938* (1938, reissued 1990) provided the ultimate chronological survey, up to the eve of our own period. Since that time the island has played a key role in a number of broader, context-setting surveys: in Richard Grove's *Green Imperialism. Colonial Expansion, Tropical Island Edens and the Origins of Environmentalism, 1600–1860* (1995), St Helena features prominently in the development of a thesis that has given birth to an entire school of thought in economic geography and environmental history; Stephen Royle's *The Company's Island. St Helena, Company Colonies and the Colonial Endeavour* (2007) forms a masterly survey of the island's fortunes during what might be called its definitive period as an outpost of the East India Company; while in *St Helena Britannica: Studies in South Atlantic Island History* (2013) Alexander Hugo Schulenburg has brought together as editor a lifetime's work by the island's most recent resident historian, Trevor W. Hearl. The island is also host to a comprehensive website with information on almost every topic of historical or current interest: http://sainthelenaisland.info/islandinformation.htm. Alongside these key sources, papers in books and learned journals dedicated to a number of scientific disciplines have added depth and texture to the historical record, among which those of Q. C. B. Cronk on botany and Reginald A. Daly on geology stand out, while *Wirebird*, the journal of the Friends of St Helena, has evolved over fifty issues into an important platform for historical as well as current debate.

Clearly there can be little to be brought to an understanding of the island's progress by a complete outsider, lacking the wealth of expertise and firsthand knowledge of this galaxy of specialists. My own attraction to St Helena was

Preface

initially an emotional and romantic one – not entirely unlike that embraced by the seventeenth century utopian writers briefly alluded to in my first chapter. Unlike them, however, the island's appeal for me lay not (or not only) in its initial, unviolated paradisical qualities, but rather in its manifestation as a microcosm of a world in whose affairs it became enmeshed, willy-nilly – an engagement from which the meagre gains accruing to the island have been offset by losses to its ecology so far-reaching as to raise doubts on any claim that it has benefited in a meaningful sense from its introduction to global society. Some exceptional figures have struggled heroically to mitigate these negative impacts, all of which ultimately were brought about by human agency. Far from its entry into European consciousness as an Edenic paradise, St Helena became both a cypher for the consequences of unfettered human despoliation of the planet and (at least potentially) a blueprint for the redemptive benefits of intelligent resource management.

Rather than merely retelling a history of these processes, the present volume aims to present no more than an introductory biography of the island – 'this little world, within itself', as Charles Darwin characterized it[1] – which nonetheless follows the successive impacts it has suffered in the course of its recorded history. The human populations implicated in these events will not be ignored, but the primary focus will be on the consequences, both lasting and transitory, of the various roles imposed on the island itself – as maritime staging post in the age of sail, as fortress, laboratory and prison. Many readers will know the island primarily for its most famous internee, Napoleon Bonaparte: he too takes his place in the text, although again with a primary emphasis on the major impact wrought on the island during his years of incarceration there.

The book is addressed to the non-specialist reader unfamiliar with the island, who may come to share something of my own interests and enthusiasms – sharpened by the advent of the Covid-19 lockdown, when contemplation of the complexities of St Helena's story provided an unfailingly appealing escape-route from the mundane world.

---

[1] Darwin 1933, 412.

Acknowledgements

Several colleagues and friends have contributed comments and insights that have been enormously valuable in compiling this study. Special thanks must go to Dr John Mcaleer, University of Southampton, Professor Donal McCracken, University of KwaZulu-Natal, Durban and Dr Henry Noltie, Royal Botanic Garden, Edinburgh, for scrutiny of the draft text and for wide-ranging and insightful commentary. M. Michel Dancoisne-Martineau, Conservateur des Domaines Français de Sainte-Hélène and honorary French consul, guided me towards what I hope may be a dispassionate and balanced account of Napoleon's incarceration on the island. Dr Vicky Heunis kindly shared with me her thesis, 'Anglo-Boereoorlog Boerekrygsgevangenekuns, 1899–1902' (University of Pretoria, 2019) – by far the most comprehensive account of the Boer prisoner-of-war camps on St Helena (and elsewhere) and of the artistic products manufactured by the detainees. Professor Jim Bennett, Keeper Emeritus at the Science Museum, London, kindly read the chapter on scientific visitors to the island. Martin Mutschlechner, historian at the Wissenschaftliche Abteilung at Schloss Schönbrunn, Vienna, provided information on Philippe Welle. Valuable help with the acquisition of images came from John Bryant; Karen Henry, Custodian, and Tracy Buckley, Assistant Custodian of Records, St Helena Archives; Cultman Collectables, Doncaster; Matthew Joshua, Head of Tourism, Treasury, Infrastructure and Sustainable Development Portfolio, St Helena Government; Dr Malgosia Nowak-Kemp, Oxford University Museum of Natural History; Karl Magee, University Archives, University of Stirling; Helena Simões Patrício, Directorate of Services, Special Collection, National Library of Portugal; Dr Andrew Pearson Rupert's Valley Archaeological Project; Edward Thorpe, W.A Thorpe & Sons Ltd., Jamestown; John Turner, site editor, sainthelenaisland.info; and John van Wyhe, director, Darwin Online.

1

Genesis

All commentators agree that the first encounter with St Helena is best conceived from sea-level: so many chapters in the island's history open from this dramatic - even intimidating - perspective. The effect is well captured in one of the earliest representations of the island, that of 1601 by the German cartographer and engraver Theodore de Bry (Figure 1): although de Bry's view (evidently compiled at second hand, for he never ventured across the Atlantic) lacks accuracy in terms of detailed topography, it succeeds in capturing something of the drama of the mariner's experience, enlarged upon by a much later map-maker, Captain Edmund Palmer, in the introduction to his *Military Sketch* of the island, compiled in 1850–52:

> St Helena lies in the strength of the s.e. trade wind, and is usually sighted by ships at a distance of 20 leagues, rising like a huge fortress from the bosom of the ocean. It is surrounded by a wall of precipitous cliffs from 1000 to 1800 feet in height, intersected by chasms, serving as an outlet for the water-courses of the island, and terminating in small coves more or less exposed to the fury of the waves. There are no less than twenty-three of these openings around the coast; but landing is almost impracticable except on the north-western or leeward side, and at Prosperous and Sandy Bays to windward, and even then only in favourable weather …[1]

Much reliance will be placed in the following chapters on similar personal impressions of the island by writers and artists (see also Plate 1). For the moment, these initial snapshots will serve to establish an image of its lowering bulk, erupting from the otherwise unbroken surface of the ocean, before something of its principal features are set out in more objective terms.

The island and its setting

The Mid-Atlantic Ridge, on which St Helena forms an outlier, runs (for the most part below sea-level) roughly from Jan Mayen Island in the north to the sub-Antarctic Bouvet Island. The ridge marks the junction of two diverging tectonic

---

[1] Palmer 1858–59, 364.

Fig. 1 *The Island of St Helena*, by the Flemish engraver Theodore de Bry (1528–1598) published in Frankfurt, 1601 as an illustration to van Linschoten's *Voyages*, part III.

Genesis

plates, pushed up from the ocean floor by the irregular escape of magma through the slowly expanding fissure, resulting at intervals in volcanoes that breach the ocean's surface. St Helena (like its nearest neighbours, Ascension Island and Tristan da Cunha) is one such volcanic island, formed some 14.5 million years ago but inert for the past 6 million years.[2] The cone rises from the sea bed at a depth of nearly 14,000 feet and continues for a further 2,700 feet above sea-level (at the highest point, Diana's Peak).[3] The island has an area of some 47 square miles (Plate 2); nearly 99 per cent of the visible mass, representing the youngest period of volcanic activity, is basaltic in character and composite in structure: the north-eastern massif forms the older area, partly overlaid by lava flows from activity in the younger, main massif. Intrusive pipes and dikes of volcanic material penetrate the cone from below – profusely so in some areas – while later erosion has fissured the surface of both eminences with canyon-like stream courses, reaching depths of nearly 1,000 feet in places.[4]

Lying around 15° 58' south and 5° 43' west, the island enjoys a benign, tropical climate: annual temperatures at coastal level vary from around 15 to 32° centigrade, some 5–6° higher than on the uplands of the interior; conversely, rainfall is much higher on the hills, reaching over 40 inches a year – about five times the coastal precipitation – giving rise to numerous streams.

The remote ocean setting of the island, some 1,100 miles to the west of southern Angola and 2,000 miles east of Brazil, combined with its (in geological terms) comparatively recent formation, demand that the whole spectrum of plant and animal life had to reach its initially barren volcanic surface from elsewhere. In the surrounding seas the dominating oceanographic system is the Benguela Current, the northward-flowing eastern arm of the South Atlantic Gyre, which is joined irregularly at the Cape of Good Hope by eddies originating in the southern Indian Ocean – a feature of importance in St Helena's biological development as well as its more recent maritime history. The current is driven by the south-easterly trade winds that dominate the regional weather system; in the era of the entry of St Helena into recorded history, these were of primary importance in carrying homeward-bound sailing vessels rounding the Cape on a course that would take them directly to the island (Plates 3–4).[5]

These same oceanographic features evidently played a crucial role in conveying the fundamental elements of a burgeoning ecosystem to the island. The process

---

[2] Continuing movement of the tectonic plates has resulted in St Helena being now located over 300 miles east of the spreading-centre of the ridge.
[3] Charles Darwin, arriving off the island on board the *Beagle* in 1836, described its appearance in some awe as rising 'like a huge castle from the ocean. A great wall, built of successive streams of black lava, forms around its whole circuit, a bold coast' (Darwin 1933, 409).
[4] Discussed most fully in Daly 1927.
[5] The same pattern of wind and tide would make it customary for outward-bound ships from Europe to adopt a westward course at the equator, crossing the Atlantic before heading south off Brazil and finally east to round the Cape.

would have been immensely slow: given the geological age of the island, it has been calculated that as little as one new colonization every 100,000 years would have been sufficient to account for the limited stock of flowering plants that developed there, whether arriving by oceanic drift or carried fortuitously by seabirds.[6] Evidence for the earliest forms of plant life to have established themselves is already long since lost due to the natural processes of erosion and decay,[7] while the present-day flora, although including a number of taxonomically isolated species, has been massively impacted during the five centuries in which the island has supported a human population. The Great Wood that densely clothed the moist, upper regions and was much praised by early mariners as a rich source of raw materials, has been wiped out by the processes of exploitation they began. Of those endemic plants that survive as relict populations, many exist only as isolated individuals – some indeed as single plants.[8] Thickets of tree-ferns (*Dickinsonia arborescens*) represent the most striking surviving feature of the early plant cover; formerly noted as growing up to 20 feet in height, the surviving population is smaller in stature and confined to the steep, upper reaches of the central ridge, where it thrives in the cool, moisture-laden atmosphere (Plates 5–6). More sparsely represented survivors from the early flora include mid-altitude Gumwood (*Commidendrum robustum*) scrub; the once-widespread St Helena Ebony (*Trochetiopsis ebenus*) was saved from oblivion as recently as 1980 and is now carefully husbanded;[9] similarly, the surviving St Helena Redwood trees (*Trochetiopsis erythroxylon*), much admired by early visitors, all now derive from a single specimen recovered on the brink of extinction in the 1950s. The island's famous Cabbage Trees are much reduced but survive in three varieties, *Senecio redivius*, *Senecio leucandron* and *Melanodendron integrifolium*. The list of endemic species discovered in intensive surveys continues to expand – many, like the grass species *Eragrostis episcopulus*, first described only in 2013, surviving merely as 'refugial pockets of native diversity'.[10]

The degree to which the flora native to the island diverges taxonomically from that of its continental neighbours is attributed to progressive evolution and extinction in those primary areas after the respective episodes of recruitment to St Helena had taken place, rather than resulting from evolutionary change operating on a

---

6   Ashmole and Ashmole 2000, 24–25.
7   Traces have been found in the fossil record, however, for the former presence of species of palm: these may have become extinct during one or other of the wetter phases detected geomorphologically (Cronk 1987, 512).
8   For a survey of the vegetation history of the island, see Cronk 1989, who gives a list and gazetteer of surviving indigenous plants.
9   For the recovery and propagation of the St Helena Ebony by the brothers Charles and George Benjamin in association with Quentin Cronk, see Chapter 6.
10   Lambdon *et al.* 2013.

significant scale within the island. The island series therefore represent the remains of a more widespread ancient flora rather than independently evolved species.[11]

The particular affinities of this relict flora are with the southern part of continental Africa, although it includes also an element suggesting a Mascarene origin. Waterborne seeds and buoyant capsules could have been carried southwards from the Mascarene Islands in the southern Indian Ocean – the hypothesis goes – by the Agulhas Current, to intersect with the Benguela Current at the southern tip of Africa and thence to make their way to St Helena.[12] Smaller seeds or fruits are likely to have been transported casually in the plumage of African birds, which turn up periodically on St Helena as vagrants or are blown there on south-easterly gales. Fossil specimens indicate the lengthy and sustained history of this additive process.[13] Some more tentative contributions from the South American flora have also been suggested.

The paucity of birds on the island today – there are no more than seven endemic species – is at least partly the result of human intervention, a much greater variety being known from the sub-fossil as well as the fossil record. In the most extensive modern survey of the St Helena bird population it was concluded by Storrs Olson that perhaps the greatest losses of species and individuals have been sustained by shearwaters and petrels (*Pterodroma* sp.), five of the six former resident species now being extinct. The island is believed once to have been perhaps 'one of the major breeding-grounds for petrels in the Atlantic', whereas now there is only a small population of a single species.[14] Other birds noted in Olson's survey of fossil remains include three forms of puffin, some with characteristics relating them to Pacific and Indian Ocean rather than Atlantic populations; these too may have died out before the arrival of a human population. Red-billed tropicbirds survive on the island today, but frigatebirds, while present in the fossil record, are now absent. Two species of flightless rails, similarly identified, are no longer present, but the date of their extinction is unestablished. Fossil remains of a large, flightless pigeon (*Dysmoropelia dekarchikos*) were found by Olson in the oldest geological deposits on the island: it too had become extinct before the arrival of the human population.[15] Other extinct species include a crake, a cuckoo and a hoopoe. The only wholly endemic species surviving nowhere else is the Wirebird (*Charadrius sanctaehelenae*) – a small plover – today limited to areas formerly occupied by

---

[11] By contrast, among the insect population, some of which undoubtedly arrived by different (airborne) means, some speciation has certainly taken place on the island (Ashmole and Ashmole 2000, 28).
[12] Alternatively, the flora now confined to the Mascarenes may itself once have been more widespread, opening up the possibility of an origin elsewhere – for example on the South African mainland. See Cronk 1987, 512.
[13] Olson 1975, 13–14.
[14] Ibid., 13.
[15] Ibid., 29.

woodland of dry gumwood, perhaps indicating its preferred habitat, although plovers are more commonly associated with more open landscapes.[16]

The insect population includes the shore woodlouse (*Littorophiloscia*) – highly likely to have arrived on driftwood in the manner described above – and a range of spiders and beetles; the latter include the She Cabbage Beetle, which lives exclusively on trees of that species and hence (like its host) is now under some pressure. Deforestation and predation or displacement by alien species have also led to great loss of snails: the Blushing Snail (*Succinea sanctaehelenae*) is the only indigenous snail still surviving, found mostly amongst the remnants of the native cloud forest.

The loss of Eden

It was with good reason that early seafarers' reports of the unspoiled St Helena were couched in terms that summoned up in European minds a veritable Garden of Eden – a utopian image applied to successive tropical island paradises that revealed themselves from the fifteenth century onwards, culminating in the great age of exploration in the Pacific in the later eighteenth century.

The island is specifically alluded to in those terms in Bishop Francis Godwin's whimsical tale of *The Man in the Moone: or a discourse of a voyage thither*, first published in 1638, in which the mythical subject, Domingo Gonsales, a native of Seville, is at death's door when he reaches the island, 'that same blessed Isle of S. Hellens, the only paradice, I thinke, that the earth yeeldeth, of the healthfullnesse of the Aire there, the fruitfullness of the soile, and the abundance of all manner of things necessary for sustaining the life of man'. The hand of man is already apparent on the island, but entirely beneficially – 'a pretty Chappell', 'divers faire walkes made by hand, and set along both sides, with fruit trees', and a landscape that 'aboundeth with Cattle, and Fowle … beyond all credit'. Having recuperated himself with all these benefits, Domingo succeeds in harnessing a number of swans that carry him off on his extra-terrestrial journey, but it could scarcely exceed the 'paradice' he left behind.[17]

While Godwin's text is reliant for details of the island on a number of sources which will form the subject of the following chapter (he even gives an illustration – see Figure 2 – as fanciful as Gonsales's swans, but with a basis in previously published topographical images), it is significant in locating St Helena not in the physical world but in the utopian cosmos in which it was imagined among Europeans of a philosophical turn of mind. Sadly, this Edenic discourse emerged just as the true impact of European contact, so romantically subsumed into Godwin's text, burst upon it with awful consequences. Within a century and a half of its discovery, the finely balanced equilibrium of the pristine garden landscape was well on its way to becoming an ecological disaster zone.

---

16   Cronk 1989, 50; McCulloch 1991.
17   Godwin 1638, 14–24. See also Hearle 1999.

Fig. 2 Frontispiece from Francis Godwin's *The Man in the Moone: or a Discourse of a Voyage thither* (1638).

2

The breach:
Europe and St Helena collide

Nova ... had a very quick and easy passage [from Calicut] to the Cape. Some time after he turned it, he discovered a little island lying in 15 degrees south latitude, to which he gave the name St Helena. This island standing by itself in the midst of such a vast ocean, seems, as if it were to have been placed there by Providence, for the reception and shelter of weather-beaten ships in their return from the Indian Ocean.[1]

The end of aeons of unfettered natural development on St Helena was signalled in 1502, when the Galician commander João da Nova (1460–1509), with a modest flotilla under his command, is said to have happened by chance on the island on his return voyage from India.[2] The year is commonly acknowledged, but narrowing the date with greater precision is problematic. The island had become widely known by its present name during the 1500s before Jan Huygen van Linschoten first asserted in his *Itinerario* of 1596 (translated into English as a *Discours of Voyages* two years later) that it was so named 'because the Portingales discovered it uppon Saint Helens day, which is the twentie one of May'.[3] The testimony of such a venerable authority became widely accepted, not least by the island's first historian, T. H. Brooke, on whose volume of 1808 subsequent generations of writers have leaned heavily.[4] Under the more sceptical gaze of recent scholarship, however, and with recognition of the disparity between the calendar of saints' days that would have been observed by the Protestant Dutch merchant van Linschoten and the (necessarily pre-Reformation) feasts celebrated in Nova's day, certainty evaporates.[5] It seems apposite that the very moment of

[1] Osorio 1752, I, 126.
[2] João da Nova's arrival on the island is now indelibly inscribed in the foundation history of St Helena: it is at least the first documented occasion when the Portuguese set foot there.
[3] Van Linschoten 1885, 254.
[4] Brooke 1808, 35.
[5] Bruce 2015, 324–46. Bruce observes that the date for St Helena's day given by van Linschoten is that which would have been celebrated in Protestant Holland at the end

The breach: Europe and St Helena collide

European contact with the South Atlantic island should be clouded by dispute, for many years of contested ownership of the island paradise lay ahead, with claim and counter-claim asserted by nations lying thousands of miles to the north.

The Portuguese, Dutch, and the British

Little more than a decade after its precipitous outline first materialized before the eyes of the lookouts at the mast-heads of the homeward-bound Portuguese fleet – and doubtless after fleeting opportune visits from other vessels of the same nation – St Helena received its first long-term settler in the form of the unfortunate Fernão Lopez, once noble and now an abject specimen after being tortured and maimed by his compatriots in Goa.[6] Lopez was abandoned on the island – willingly, it would appear – along with some food, cooking-pots and clothing. Later crews touching St Helena left not only food but also fruit and vegetables that he might plant, 'so that he cultivated a great many gourds, pomegranates, and palm trees, and kept ducks, hens, sows, and she-goats with young, all of which increased largely, and all became wild in the wood'.[7] Ultimately Lopez's presence on the island was officially sanctioned by the Portuguese monarch;[8] he now became more open with visiting ships, providing them with the produce he grew so successfully until finally he died in 1546.

---

of the sixteenth century, but that da Nova would undoubtedly have observed the day prescribed by the Catholic calendar, on 18 August. Da Nova's fleet went on to reach Lisbon on 11 September – an impossible undertaking in little more than three weeks – rendering the 18 August date equally problematic, but Bruce adduces a third possibility: Odoardo Duarte Lopes, who visited the island in 1578, claimed that the Portuguese had discovered the island 'on the 3rd of May, the Feast of St Helena' – again an error, but corresponding to the Feast of the True Cross, as discovered by Helena in Jerusalem. As early as 1638 Sir Thomas Herbert also favoured 3 May, 'a day consecrated to the memory of Helena the Empress first found the Crosse' (Herbert 1677, 391).

[6] While resident in Goa, Lopez had unwisely renounced his religion and joined the ruler of Calicut in arms against the Portuguese. When a reconquest under Afonso de Albuquerque reasserted Portuguese supremacy, Lopez narrowly avoided execution but along with a number of other renegade apostates was subjected to appalling mutilation in retribution 'for the treason and wickedness which they had committed against God and their King': his nose, ears, right hand and left thumb were cut off and his prolonged persecution continued until he managed to board a ship bound for Portugal. When the vessel called at St Helena to take on fresh water, Lopez sought refuge on the island and evaded a search by the crew (who evidently were not ill-disposed towards him).

[7] Birch 1880, xxxvi–xxxix, 238–40.

[8] Manuel I was so intrigued as to summon Lopez to Lisbon; a nocturnal audience with the king and queen took place at which he was offered a hermitage in which to live out his days in seclusion, but instead he asked to be allowed to seek absolution from the Pope and to return to St Helena, all of which came to pass.

The privilege of residence granted to Lopez was an exceptional one, although other 'eremites' followed in due course, charged with caring for the sick sailors left from time to time to recuperate.⁹ A woodcut from van Linschoten's *Discours of Voyages* (1598) shows what was then known to the English as Chapel Valley, as yet occupied only by the chapel built by order of the king of Portugal and dedicated to St Helena¹⁰ (see Figure 3), but even the 'eremites' mentioned above could occasionally be tempted to exploit their privileged position. Van Linschoten, for example, writes of one of them, a Franciscan:

> In time past there dwelt an Hermet in the Ile, [who continued there for] certaine yeares, under pretence of doing penance, and to uphold the Church, hee killed many of the Goates and Buckes, so that everie year hee sold at least five or six hundred skinnes, and made great profit thereof: which the King hearing, caused him presently to bee brought from thence into Portingall.¹¹

His fate thereafter is unrecorded.

João da Nova's discovery of St Helena had taken place within five years of the first rounding of the Cape by Vasco da Gama, which opened Portuguese contact with India. The remainder of the sixteenth century saw the launching of regular voyages for the purposes of trade, particularly through the Portuguese colonies established on the sub-continent. Throughout this time the island played a fundamental role in the success of these ventures – specifically on the return voyage when, having rounded the Cape, vessels would be carried by the trade winds on a steady north-westerly course (see Plate 4).¹² Frequently their crews would already have suffered considerable privation on the passage from India, so that we can well believe the enthusiastic assessment of the island's strategic importance given by van Linschoten, who sailed as a pilot with the Portuguese for a number of years:

> it is an earthly Paradise for ye Portingall shippes, and seemeth to have been miraculously discovered for the refreshing and service of the same, considering the smalnesse and highnesse of the land, lying in the middle of the Ocean seas, and so far from the firme land or any other Ilands, that it seemeth to be a Boye, placed in the midle of the Spanish Sea[s]: for if this Iland were not, it were impossible for the shippes to make any good or prosperous Viage: for it hath often fallen out, that some shippes which have missed thereof, have indured the greatest miserie in ye world, and were forced to put in to the coast

---

9   Van Linschoten 1885, 257. The Portuguese monarchy was also said to have been fearful that anyone granted residence might begin to harbour thoughts of independence.
10 Hakluyt 1903–05, VI, 34.
11 Van Linschoten 1885, 257.
12 A late sixteenth-century Portuguese mariner's account of the island mentions the impossibility of setting a course due south for St Helena, 'because there is always a headwind; and you have to go down to at least 16 degrees – which is the latitude of [St Helena] – before it is possible to steer west' (reproduced in Cronk 2000, 98).

The breach: Europe and St Helena collide

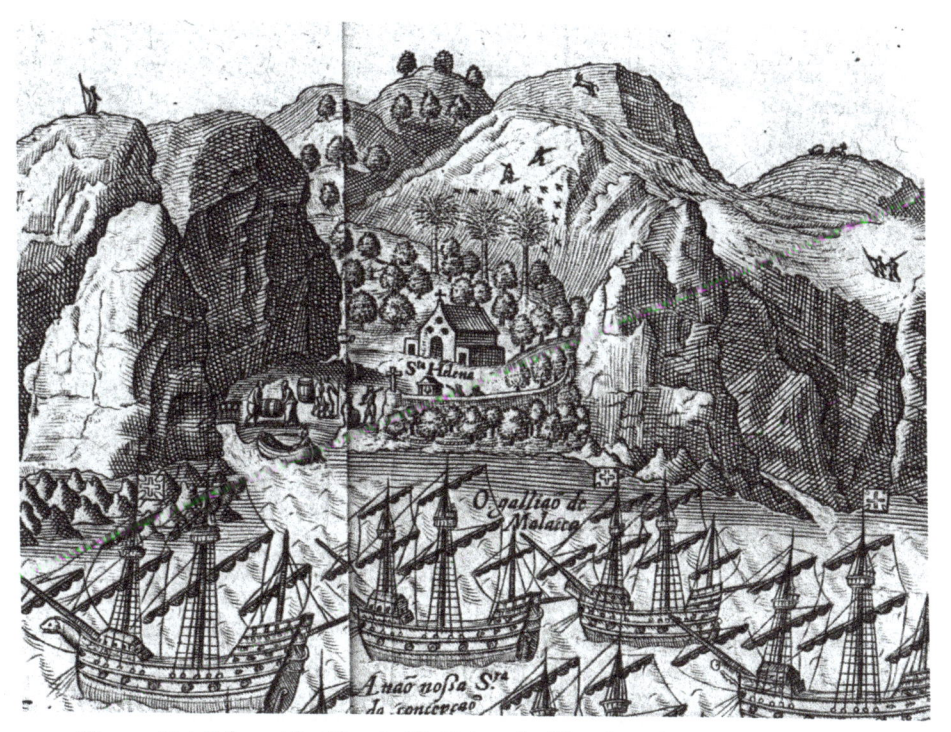

Fig. 3   Detail from *The Island of St Helena*, by Theodore de Bry (1601) showing Chapel Valley (later James Valley), as yet without permanent settlement apart from the eponymous chapel. Casks of fresh water are being loaded by the fleet assembled for the final leg of the homeward voyage.

of Guinea, there to stay [until] the falling of the raine, and so to get fresh water, and afterwardes came halfe dead and spoyled into Portingall.[13]

Conscious of these tremendous advantages, the Portuguese no doubt did their best to keep the location of St Helena a secret, but by the second half of the sixteenth century it had evidently become more widely known. In 1582–83 Edward Fenton, commissioned by the earl of Leicester to lead a squadron of four ships with a view to establishing trade with China, determined instead to divert to St Helena, seize any Portuguese ships he might find there and establish himself as ruler of the island.[14] Although the venture came to nothing, it serves to demonstrate that knowledge of the island and of its strategic importance to the developing Far Eastern trade route was already in circulation before the better-documented arrival a few years later of

---

[13]   Van Linschoten 1885, 256.
[14]   Fenton 1957. Fenton's editor, E. G. R. Taylor, observes (p. xliv) that the scheme 'was not quite so ridiculous as it sounds', and had it been followed through would have stood some chance of success.

the privateer and explorer Captain Thomas Cavendish. On the morning of 8 June 1588 Cavendish stumbled upon St Helena by chance on the return leg of a voyage that had taken him round the world and on which he had grown fabulously wealthy at the expense of the Spanish on the west coast of the Americas. The encounter must have been an extraordinary one for Cavendish, for when he and his crew went ashore the following afternoon they found the island deserted but with clear signs of a former European presence, which he evidently recognized as Portuguese:

> ... wee found a marvellous faire & pleasant valley, wherein divers handsome buildings and houses were set up; and especially one which was a Church,[15] which was tiled & whited on the outside very faire, and made with a porch, and within the Church at the upper end was set an altar, whereon stood a very large table, set in a frame having in it the picture of our Saviour Christ upon the Crosse, and the image of our Lady praying, with divers other histories curiously painted in the same. The sides of the Church were all hanged with stained clothes having many devices drawen in them ...
>
> There are two houses adjoyning to the Church, on each side one, which serve for kitchins to dresse meate in, with necessary roomes and houses of office ... and through both the saide houses runneth a very good and holsome streame of fresh water.
>
> There is also right over against the saide Church a faire causey made up with stones reaching unto a valley by the seaside, in which is planted a garden, wherein grow great store of pompions and melons: And upon the saide causey is a frame erected whereon hange two bells wherewith they ring to Masse; and hard unto it is a Crosse set up, which is squared, framed and made very artificially of free stone, wheron is carved in cyphers what time it was builded, which was in the yeere of Our Lord 1571.[16]

As the English shortly began probing the possibility of trade via the Cape with the East Indies, they too were quick to avail themselves of the island's conveniences – not only in terms of food and water but also for the refuge it offered

---

[15] The chronicler of Jacob Roggeveen's voyage to the Pacific in the 1720s records that the chapel was first built from the timbers of a Portuguese ship that had been wrecked on the island (Callander 1768, 524). It was apposite that the fabric of the island's first building derived from the shipping that proved integral to St Helena's growing importance.

[16] Cavendish's description is taken from Hakluyt 1903–05, XI, 343–44. Two years before Cavendish's encounter with the island, Guido Gualtieri had already published an account of a voyage which carried four young Japanese 'ambassadors' to Europe at the behest of the Jesuits, in the course of which they touched at the island; he lists the invaluable store of animal and vegetable resources and abundant water supplies that proved a salvation to homeward-bound mariners (Gualtieri 1586, 48–49). Knowledge of the island therefore was already beginning to be disseminated amongst the European maritime nations.

The breach: Europe and St Helena collide

for the stream of sailors frequently brought to death's door by the rigours of the homeward voyage and the hazards of a diet almost totally bereft of fresh fruit, green vegetables or meat. Another early visitor, Sir James Lancaster, arriving on 3 April 1593, was surprised to find one of his countrymen who had been left there in exactly those circumstances. When Lancaster's surgeon, Edmund Barker, went ashore, he recorded that:

> in an house by the chappell I found an Englishman, one John Segar of Burie in Suffolke, who was left there eighteene moneths before by Abraham Kendall, who put in there with the *Roiall Marchant*, and left hime there to refresh him on the iland, being otherwise like to have perished on shipboord; and at our comming wee found him as fresh in colour and in as good plight of body to our seeming as it might be, but crazed in minde and halfe out of his wits, as afterward wee perceived.[17]

The vessel mentioned by Barker had been one of three that sailed from England, bound for India, under Lancaster in 1591. By the time they reached the Cape, so many of the crews had come down with scurvy that the sick had been loaded onto the *Royal Merchant* and sent back to England under Captain Kendall. It is a measure of the efficacy of St Helena's anti-scorbutic herbs, its resources in foodstuffs, and the island's healthful climate that of the fifty men Kendall landed there, all except Segar had made a rapid recovery.[18]

As well as English ships, Spanish, French and especially Dutch vessels now habitually called at the island for the purposes of refreshing themselves. St Helena remained uninhabited at this time, save for occasional convalescent mariners. From time to time more exotic and unexpected visitors put in an appearance, such as the Patriarch of Abyssinia – an awkward character by all accounts, who insisted on being left behind there in the course of a voyage to Portugal; he stayed for a year, starved of enjoyment and 'suffering some corporal discomfort from hunger and other needs'.[19] In the valley around the chapel there now stood some thirty or forty dwellings which would be occupied whenever fleets gathered there to form convoys for greater security on the last stretch of the homeward voyage. At times the crews would be so numerous that the seamen had to camp out in the surroundings, as recorded by van Linschoten:

---

[17] Having regained his physical health, Segar evidently remained in a fragile mental state. Barker continues: 'whether he were put in fright of us, not knowing at first what we were, whether friends or foes, or of sudden joy when he understood we were his old consorts and countreymen, hee became idle-headed, and for eight dayes space neither night nor day tooke any naturall rest, and so at length died for lacke of sleepe' (Lancaster 1940, 16, 25).

[18] Gosse 1990, 22–23.

[19] Ibid., 12.

When the ships come thether, everie man maketh his lodging under a tree, setting a Tent about it: for that the trees are so thicke, that it presently seemeth a little towne or an armie lying in the fielde. Everie man provideth for himselfe, both flesh, fish, fruit, and woode, for there is enough for them all: and everie man washeth [his] Linnen …

There they use [sometimes] to Carve their names, and markes in trees & plants for a perpetuall memorie: whereof many hundreth are there to be found, which letters with the growing of the trees, doe also grow bigger and bigger, we found names that had been there since the yeare of the Lord 1510 & 1515, and everie yeare [orderly] following, which names stoode upon Figge trees, every letter being of the bignesse of a spanne, by reason of the age and growing of the trees.[20]

While contact remained uncompetitive, relations between the nations frequenting the island could be amicable when not strained by tensions in Europe. Arriving at the island in February 1603, for example, Sir James Lancaster records that 'wee delivered unto the Frenchmen and unto the Hollanders such victualles to relieve them as we could spare; which was six hogsheades of porke, two [hundredweight] of stockfish, one hogshead of beanes, and five hundred of bread, wherof the Hollanders were in great want.'[21] In the course of time, however, the Portuguese – who had in the meantime established several footholds on the West African coast – found themselves increasingly under pressure at St Helena and their fully-laden carracks liable to attack from British privateers and more commonly from the Dutch. These deteriorating relations were recorded on the face of the island, where the whitewashed chapel that had stood for several generations fell victim to the puritan ire of Protestant iconoclasts and its free-standing stone cross was shattered. When Sir Thomas Herbert visited the island in 1629 he found only 'some ruins of a little town', whose destruction he attributed to the Spanish, while the chapel was described as 'by the Dutch of late pulled down, a place once intended for God's worship, but now disposed of to common uses'.[22] Even the natural resources that had been a boon to sailors of all nations were targeted in an increasingly bitter process of attrition that began with the opening of the seventeenth century. The descent from a state of grace is well recorded in the journal of the French navigator François Pyrard, who visited twice, in 1601 and 1610 (having spent six of the intervening years as a prisoner on the Maldives). The decline of the chapel, which he had first encountered as 'adorned with a fair altar and handsome images and pictures', with 'a fine large cross of freestone, white as marble and well carved' standing before it, he attributed to mutual hostility between the Portuguese and the Dutch, who had also taken to destroying each other's messages left for their respective compatriots. The fruit trees, which had

[20] Van Linschoten 1885, 258.
[21] Lancaster 1940, 139.
[22] Herbert 1677, 392.

The breach: Europe and St Helena collide

been such a boon to visiting vessels, had been extensively cut down, ostensibly by the Portuguese, so as to deny their benefits to the Dutch and the English.[23] Occasionally more direct confrontations took place, in one of which in 1613, two Portuguese carracks succeeded in sending the 700-ton Dutch East Indiaman *Witte Leeuw*, on her homeward voyage from Bantam, to the bottom of Jamestown Bay: seemingly due to the misfiring of one of her own cannon, her powder magazine ignited causing a catastrophic explosion that sank her within minutes. Part of the *Witte Leeuw*'s cargo, which included peppercorns, nutmeg, cloves and porcelain, was recovered by divers led by Robert Sténuit in 1976 (Plate 7).[24] Together with the surviving ship's manifest, the wreck provides eloquent evidence of the centrality of St Helena to the growing contact between Europe and the Far East at this time and of the way in which European political, religious and mercantile tensions continued to extend their influence into the remote South Atlantic.

For a brief period the Dutch held sway on the island. An extraordinarily rare surviving proclamation confirms their formal claim to the territory (Plate 8). It reads in translation:

> On the 15th day of April 1633 the noble sire Jacques Specx, late Governor General of the State of the United Provinces of the Netherlands in India, together with the Council-in-pleno of the Dutch fleet which has just arrived here ... have accepted the possession and proprietorship of the island, named of yore St Helena, with all its grounds, hills, cliffs and rocks belonging to it, for the State of the United Provinces, in order to the benefit and advantage of the said Netherland State, as soon as the circumstances shall allow, to fortify, occupy, populate and defend it against the invasion of enemies, in the way as their Highnesses the High and Mighty States General of the said United Provinces shall deem advisable ...
>
> To certify this and confirming the truth, that nobody may pretend ignorance thereof has been erected this pillar, as well as this notification, duly sealed

---

[23] Pyrard 1890, 296–302. The altar and the cross recorded by Pyrard are themselves thought to have replaced those 'beate downe' by Cavendish's crew in 1588 (p. 297 note 2).

[24] An account of the salvage operation is given by the expedition's leader, Robert Sténuit, at http://www.thunting.com/smf/shipwrecks_sunken_treasure/witte_leeuw_voc_shipwreck-t8058.0.html. While much of the organic component of the cargo had perished, the contents of over 15,000 bags of peppercorns lay consolidated on the seabed in a deposit up to 6 feet thick in places. Most spectacular, however, were the ceramics recovered: a mass of Jingdezhen export wares (termed kraakwares) – originating in China but widely traded to South-East Asia – as well as martaban jars from as-yet ill-defined production centres (perhaps in Thailand or Burma); a number of European stoneware flagons in use by the crew were also recovered. Most of these finds are now held in the Rijksmuseum, Amsterdam. The story of the ship and its cargo is examined in detail by van der Pijl-Ketel (1982). Later divers on the wreck site (including Jacques Cousteau in the 1970s) have recovered a further cannon (bringing the total salvaged to eight) and other items.

and signed and nailed thereunto in the above mentioned year and on the date mentioned.[25]

There is, however, no evidence that the Dutch ever succeeded in implementing their intention to occupy or fortify or even to populate the island (and the repatriation of the document almost undamaged appears to suggest that the claim was not pursued at length); by mid-century the superior attractions of the mainland colony they established at the Cape came to fulfil the needs of the VOC's fleet as a way-station to and from the East Indies.[26] Now it was the turn of the British, in the form of the East India Company, to lay claim to the island. Apart from a brief interlude when the Dutch sought almost casually to reassert their claim (Chapter 5), St Helena would remain henceforth a British possession.

---

[25] Translation from Gosse 1990, 41.
[26] The claim that St Helena had for a time been a Dutch possession was deemed by Sir William Foster (1919) to have had no basis, in view of their having failed to establish a resident population on the island.

3
Population and environment: early impacts

The earliest descriptions of the island, compiled by mariners for whom it represented (often quite literally) a salvation, tend to open with an impression of its forbidding appearance from the sea followed by a progress up the well-watered Chapel Valley and the unfolding of the upland landscape into greenery. Peter Mundy, for example, observed in 1634 that:

> The Island is verie pleasant to see to, alofte in some places faire woods of small Trees with straight stemms and broad bushey Topps, and in other places of other sorts; fine round, smooth hills with excellent grasse; many thicketts of Ferne, etts. runninge water in the bottomes [of the vallies] etts. and groves of trees ...[1]

By Mundy's day, however, the face of the island had already undergone far-reaching changes, directly or indirectly, at the hands of transitory seamen. Timber was a major requirement of the ships that touched there, many of which found themselves in immediate need of repairs – sometimes on a major scale – in order to complete the long voyage home from the East Indies. Native redwood in particular, once widespread on the island, was sought out by ships' carpenters desperate for materials to see them through the remainder of the voyage to England.[2]

With the permanent settlement of the island in the mid-seventeenth century (Chapter 4), consumption of timber for construction increased markedly and pressure on the forests rose in line with the growing demand for firewood. Domestic heating and cooking accounted for an increasing demand, but other factors are also singled out as having been particularly destructive. Yams, introduced particularly to feed the enslaved population (see below), required prolonged boiling over several hours to make them at all palatable, while the distillation of arak, to which large numbers of the white planters quickly became addicted, also consumed significant quantities of timber. With the passage of time, so much of the immediately available timber had been cleared that those gathering firewood (again slaves in particular) were having to walk for several hours into the interior to find new sources.

---

[1] Mundy 1914, 330.
[2] Early reference to certain trees by the term 'sparwood' indicates all too clearly their harvesting for the replacement of damaged timbers.

While there is no documented instance of deliberate clearance of woodland to make way for agriculture, one pernicious factor ensured that once trees had been felled the woodland would be lost forever – namely, the presence of numerous feral goats and other ruminants. Casually introduced by the Portuguese from the early part of the sixteenth century, goats succeeded spectacularly well in establishing themselves in the wild. Although they provided a welcome source of fresh meat to visiting mariners, the effect of goats on limiting the capacity of the forests to regenerate was catastrophic: naturally-seeded saplings thrown out by mature trees proved irresistible to them, and every time a tree was lost by felling or due to old age, they ensured that it would produce no successors. The tall trees of the mature forests so admired by early visitors were immune to the attentions of the goats, but when these reached the end of their lives there would be no new growth;[3] the consequent loss of habitat proved equally damaging to all the associated herbaceous plants, birds and insects.

By the end of the seventeenth century, losses to the island's environment were no longer to be measured in individual species or even in whole woods or groves. So advanced was the process of deforestation that the consequences came to be written on the landscape: extensive erosion followed inexorably from the removal of trees with their capacity to store water and to consolidate the soil. Gullies and crevasses opened up in response to the torrents of water that flowed regularly in the absence of the buffer-zone formerly established by tree cover. The island had begun its long slide into ecological crisis.

Animal and botanical resources

Although early human interaction with the flora and fauna was largely negative in its impact, introductions outnumbered extinctions in the earliest contact period, although these could prove equally devastating in their effect on native species.

Perhaps the earliest alien introductions are those conventionally attributed to the presence of the island's first resident European, Fernão Lopez. In the years following Lopez's self-imposed exile there (Chapter 2), passing Portuguese vessels took care to ensure that he did not starve, and of the plants and animals he raised there it was said that 'all became wild in the wood'.[4]

---

[3] Charles Darwin (1890, 469) mused on this phenomenon: 'The fact that the goats and hogs destroyed all the young trees as they sprang up, and that in the course of time the old ones, which were safe from their attacks, perished from age, seems clearly made out … It is very interesting thus to find, that the arrival of animals at St Helena … did not change the whole aspect of the island, until a period of two hundred and twenty years had elapsed: for the goats were introduced in 1502, and in 1724 it is said "the old trees had mostly fallen".' The author of Roggeveen's chronicle comments on the trees surviving in the 1720s that there were 'none fit for timber, but for fuel only' (Callander 1768, 524).

[4] Birch 1880, xxxvii.

Population and environment: early impacts

No doubt an element of self-interest also motivated these periodic depositions, but in any case by the time of Thomas Cavendish's arrival in 1588 the lower reaches of Chapel Valley (today with Jamestown at its foot) already presented a highly modified landscape:

> This valley is the fairest and largest lowe plot in all the yland, and it is marveilous sweete and pleasant, and planted in every place either with fruite trees, or with herbes. There are fig trees, which beare fruit continually, & marveilous plentifully: for on every tree you shal have blossoms, greene figs, and ripe figs, all at ones: and it is so all the yere long: the reason is that the yland standeth so neere the Sunne. There be also great store of lymon trees, orange trees, pomegranate trees, pome-citron trees, date trees, which beare fruite as the fig trees do, and are planted carefully and very artificially with very pleasant walkes under and betweene them, and the saide walkes bee overshadowed with the leaves of the trees: and in every voyde place is planted parceley, sorrell, basill, fenell, annis seede, mustard seede, radishes, and many speciall good herbes: and the fresh water brooke runneth through divers places of this orchard, and may with very small paines be made to water any one tree in the valley …[5]

Virtually every one of these plants was an introduced species, and evidently they had already started to spread, for Mundy continues:

> The yland is altogether high mountaines and steepe valleis, except it be in the tops of some hilles, and downe below in some of the valleis, where marveilous store of all these kinds of fruits before spoken of do grow: there is greater store growing in the tops of the mountaines than below in the valleis: but it is wonderful laboursome and also dangerous travelling up unto them and downe again, by reason of the height and steepnesse of the hilles.[6]

François Pyrard's assessment (Chapter 2) of the negative effects of the destructive dispute festering between the Portuguese and the Dutch (if they were indeed solely to blame) was corroborated in 1649 by his countryman Jean-Baptiste Tavernier:

> There are great store of Citrons, and some Oranges, which the Portugals had formerly planted there. For that Nation has that vertue, that wherever they come, they make the place the better for those that come after them; whereas the Hollanders endeavour to destroy all things, wherever they set footing. I confess

---

[5] Hakluyt 1903–05, XI, 344. Tobacco is also mentioned as a Portuguese introduction by Jon Ólafsson, a gunner on the Danish ship *Pearl*, in 1625 (Ólafsson 1932, II, 206); Joseph Banks later concurred, mentioning that the Portuguese, 'finding this herb particularly beneficial in complaints contracted in long voyages made a point of sewing it wherever they went ashore, a custom from whence all nations have since reapd no small benefit' (Banks 1962, II, 268). Peter Mundy later found growing among the rocks 'some mints, malloes [and] purcelane', as well as 'a kind of camomile smelling sweet': the latter (*Cotula anthemoides*), at least, is said to be native to the island (Gosse 1990, 31).

[6] Gosse 1990, 345.

the Commanders are not of that humour, but the Sea-men and Souldiers, who cry one to another, we shall never come hither any more, and out of greediness will cut down a whole tree instead of gathering the fruit.[7]

Perhaps unsurprisingly, this wanton destruction was especially evident in the Chapel Valley area, where the Portuguese in particular had made a sustained effort at husbanding the gardens and lemon groves they established but which languished unattended after the Portuguese withdrawal from the island. When the East India Company (EIC) commander Thomas Best called in 1614 he found no lemons were to be had in that area, though when he sent his boats westward to what had already become known as Lemon Valley his men were able to gather them by the thousand.[8] Twenty years later Mundy estimated that there were only some forty lemon trees left on the whole island, twenty in Lemon Valley and the remainder scattered in small clusters, though on another visit he found among the woods 'other trees Not Formerly knowne by them, Most bending with their burthens, on whome besides the Multitude off well coulloured ripe ones were as Many greene and smalle, and Many More blossomes'.[9]

The island maintained, meanwhile, its reputation as a sanctuary for sailors afflicted with 'scorbutick distempers', whose 'only hopes are to get refreshment and health' here. The words are those of William Dampier, whose ship *Defence* anchored at St Helena for a few days in 1691. He continues:

For the Islands afford abundance of delicate Herbs, wherewith the Sick are first bathed, to supple their Joints, and then the Fruits and Herbs, and fresh food soon after cure them of their scorbutick Humours. So that in a Week's time Men that have been carried ashore in Hammocks, and they who were wholly unable to go, have soon been able to leap and dance.[10]

Once the decision had been taken by the EIC to take control of the island, the British too began to import plants with some deliberation. Along with his commission as the first governor, Captain Dutton received the following detailed instructions from the Court of Directors:

---

[7] Tavernier 1678, part II, 207.
[8] Thirty hogs were also killed on this occasion, though Best comments that 'if we had laid ourselves out for the purpose, I daresay we might have got two hundred hogs besides many goats' (quoted in Gosse 1990, 29).
[9] Ólafsson 1932, III, part 2, 412.
[10] Dampier 1937, 365. These sentiments were confirmed a decade or so after Dampier's visit by Francis Rogers, a merchant, having just completed a voyage of over three months from Bencoolen. The island, he writes, 'is reckoned one of the most healthful places in the world, and our sailors just dead with the scurvy and other diseases, with eating nothing but salt provisions … when carried ashore, recover to a miracle, rarely any dying though never so ill when brought ashore' (Rogers 1936, 192). McCracken (2022, 15) confirms that by the late 1740s the native lemons were all but gone.

Population and environment: early impacts

Take notice that wee have ordered Capt. Robert Bowen[11] to touch at St Jago, or some one of the islands of Cape de Verde, in his outward passage. where being, wee desire your care and dilligence as speedilie as possibly may be to procure all manner of plants, rootes, graines, and all other things necessarie for plantation there to be had or procured, but more espetially in those which are to be esteemed your most certaine provisions, as planton rootes, cassada-sticks, large jamooes, potatoes and bonavist, pease, gravances, and beanes of all sortes, oranges and lemons.[12]

Imported wildfowl similarly flourished in the amenable climate of St Helena, alongside other native species. Cavendish had been pleasantly surprised in 1588 by the abundance of game birds to be found there – many of which evidently had been introduced by the Portuguese:

There is also upon this yland great store of partridges, which are very tame, not making any great hast to flie away though one come very neere tham, but onely to runne away, and get up into the steepe cliffes: we killed some of them with a fowling piece. They differ very much from our partridges which are in England both in bignesse and also in colour. For they be within a little as bigge as an henne, and are of an ashe colour, and live in covies twelve, sixteen, and twentie together: you cannot go ten or twelve score but you shall see or spring one or two covies at the least.

There are likewise no lesse store of fesants in the yland, which are also marveilous bigge and fat, surpassing those which are in our countrey in bignesse and in numbers of a company. They differ not very much in colour from the partridges before spoken of.

Wee found moreover in this place great store of Guinie cocks, which we call Turkies, of colour blacke and white, with red heads: they are much about the same bignesse which ours be of in England: their egges be white, and as bigge as a Turkies egge.[13]

Pigeons were certainly introduced, seemingly brought to the island during the sixteenth century: van Linschoten reports 'Hennes, Partridges, and Doves, by thousands' by 1589,[14] while Peter Mundy found there 'store of little speckled guinney Henns, partridges and Pigeons' in 1634.[15]

---

[11] Master of the *London*, on which Dutton travelled to take up his appointment.
[12] Foster 1919, 285. Foster identifies the more obscure plants respectively as cassava, yams, pulses and chick-peas.
[13] Hakluyt 1903–05, XI, 345–46.
[14] Van Linschoten 1885, 255.
[15] Mundy 1914, II, 330. Later George Forster would record in 1775 the presence of 'rice birds, commonly called paddies (*Loxia oryzivora*) which have been introduced from the East Indies' (Forster 2000, II, 666).

Perhaps the only truly endemic bird, 'A small land fowl, somewhat like a lark in colour, shape and flight' noted by Mundy is today identified as the plover-like *Charadrius sanctaehelenae*, known locally as the Wirebird (Plate 9): a spindly-legged inhabitant of the low-altitude barren plains and seashores, it is found nowhere but on the island.[16] Considering the numbers of feral cats and dogs referred to by various visitors in the early 1600s – not to mention the rats that later reached plague proportions – the survival of the species seems little short of miraculous. Otherwise, the only land birds that caught the attention of early visitors were the domestic fowls mentioned above.

Seabirds too provided a valuable source of protein, but were perhaps less at risk from human intruders. Mundy and his party landed one day on an offshore island (perhaps the presently-named Egg Island) where, armed with sticks, they took nearly 100 'Sea fowle, russet Coulour, almost as bigg as a pidgeon, but tast very fishey'.[17] When the EIC took control of the island, the Company asserted its right of ownership over the eggs, although the islanders were allowed to collect them on certain days of the week. Evidently this harvesting of eggs got out of hand, to the extent that Egg Island was abandoned by the terns that had nested there in favour of the more inaccessible cliffs of the seashore. An interesting record survives from 1727 of a further attempt by the Company to encourage the settlement of a potentially useful bird species:

> ... there being several birds of a different species from those that frequent the island, lately come hither, the bodies of which are as large as a pheasant, their legs long and black but their claws open and not webbed like a sea-fowl, with long bills resembling those of a snipe ... which probably may breed here if not destroyed or disturbed ... All persons be publickly forbid by advertisement either to kill or disturb any of the said birds or destroy any of their egge.[18]

More impressively numerous (and more valuable to visiting mariners) were the feral pigs and goats already mentioned that had established themselves in the wild within a few decades of their first introduction (Figure 4). Cavendish's description of the island from 1588 continues in the following vein:

> There are in this yland thousands of goates, which the Spaniards call Cabritos, which are verie wilde: you shall see one or two hundred of them together, and sometimes you may beholde them going in a flocke almost a mile long. Some of them, (whether it be the nature of the breed of them, or of the country I wot not) are as big as an asse, with a maine like an horse and a beard hanging downe

---

[16] The Wirebird is St Helena's national bird and features on the island's armorial shield.
[17] Mundy 1914, II, 332. Gosse observes that to this day the island is home to the same Noddy terns (*Anous stolidus*).
[18] East India Company Records, reproduced by Ashmole and Ashmole (2000, 27), who suggest that the species in question may have been the migratory glossy ibis (*Plegadis falcinellus*).

Fig. 4 Portuguese mariners shooting birds and goats: detail from an unattributed Portuguese map of St Helena (1600s). For the whole map see http://id.bnportugal.gov.pt/bib/catbnp/269861. Biblioteca Nacional de Portugal.

to the very ground: they will clime up the cliffes which are so steepe that a man would thinke it a thing unpossible for any living thing to goe there. We tooke and killed many of them for all their swiftnes: for there be thousands of them upon the mountaines.

Here are in like maner great store of swine which be very wilde and very fat, and of a marveilous bignes: they keepe altogether upon the mountaines, and will very seldome abide any man to come neere tham, except it be by meere chance when they be found asleepe, or otherwise, according to their kinde, be taken layed in the mire.[19]

In 1603 this population caught the attention of Sir James Lancaster: they were plentiful, he acknowledged, but 'hard to come by, unlesse good direction be given for the getting of them'. The following stratagem was adopted by Lancaster's military commander for capturing them in bulk:

He appointed foure lusty men and of the best shot he had, to goe into the iland and make their abode in the middest of it; and to every shot he appointed foure men to attend him, to carrie the goats that hee killed to the rendezvous. Thither went everie day twentie men, to fetch home to the ships what was killed ... And by this meanes the ships were plentifully relieved, and every man contented ... All our sicke men recovered their health, through the store of goats and hogs wherewith wee had refreshed ourselves.[20]

François Pyrard records that the animal population gradually got wise to the dangers posed by visiting seamen, so that

when they see the ships come in they all go off to the mountains, and when they see them depart they return to the valleys, and especially to that where the chapel is, which is the fairest and most spacious ground, and has always some plants growing, which they come to eat.

Those mariners left behind on the island to recover their health in turn adapted to this practice and developed more passive methods of capture:

The men that are left to sojourn there then catch these animals in this crafty wise: these gardens are enclosed with walls and doors, which they leave open, and, when the animals have entered, a man concealed at a distance draws a cord fastened to the door and shuts them in: thus they catch as many as they please and let the rest go.[21]

---

[19] Hakluyt 1903–05, XI, 345–46.
[20] Lancaster 1940, 119–20. When Lancaster's men had discovered the marooned Englishman John Segar on an earlier visit to the island (ibid., 25), they found Segar had dried (salted?) forty goats for his own consumption.
[21] Pyrard 1890, 301.

But for all the benefits brought by the goats (described by Mundy as 'For the most part blacke, some white and party colloured') to visiting mariners,[22] these animals would prove in the long run to have the most negative impact by far on the island's ecology. While the EIC at first extended its protection to the goats, decreeing in 1678 that 'noe person whatsoever on the said Island doe presume on any pretence whatsoever to Hunt, Kill, or Destroy any of the Said Wild Goats without leave first had in Writing from ye Gov[erno]r', by 1698 it was recognized that there had by now developed such a surfeit of both wild and domesticated animals 'that this great plenty made them neglect pounding of them: and consequently the most part thereof did runn wild'. Although the Company recognized the need for effective fencing to protect the island's dwindling timber resources, its attitude and that of the local populace towards the goats remained equivocal throughout the following century.[23]

As for the feral swine ('grizzled or grey, with very long bristles and hair'), in 1625 Jon Ólafsson described 'great herds' of them on the uplands, which could only be caught by hunting them with large numbers of men: 'They have weapons to attack them, which they call cross-lances', he writes, 'and it is the seaman's custom that any who should use them ... should leave their weapons behind on the path where it first rises towards the mountain, and anyone may use them who has need of them.'[24]

Mundy judged the island 'a most excellent place for increase of cattle', seemingly using the term in the present-day sense rather than signifying livestock in general as was customary in his day;[25] he also noted the presence in Chapel Valley of four horses, 'doubtless left here by the Hollanders to increase for provision'; de Rennefort recorded that by 1666 the English too had imported horses to the island, 'but these had become so wild that when they were pursued to the ends of the island they threw themselves off the rocks into the sea rather than be caught'.[26] Live sheep and deer are also reported to have been taken to St Helena from England: the presence of the latter may seem unlikely, but the practice of moving quite large herds of deer over hundreds of miles distance had been perfected under the Tudor and Stuart monarchies with a view to improving local populations for the hunt.[27]

Three further introduced species combined to produce a devastating effect on the island's natural wildlife. Cats and dogs, arriving as pets in some guise, soon

---

[22] Ólafsson 1932, III, part 2, 413.
[23] See further Chapter 4.
[24] Ólafsson 1932, II, 206–07. On that occasion Ólafsson's party killed nine large swine, 'but with great trouble'. The 'cross-lances' he mentions were doubtless identical with the heavy boar-spears favoured at the time in Europe, which had a cross-bar at the bottom of the blade to prevent them becoming too deeply embedded in the animal (MacGregor 2012, 124).
[25] Gosse (1990, 43) suggests that some of these, at least, may have come from two Portuguese ships wrecked at the island in 1643, 'known to have replenished the island with cattle, hogs and goats'.
[26] Gosse 1990, 54.
[27] MacGregor 1992.

established feral populations that wrought havoc amongst the bird population in particular. By 1717 the island was home to 'vast numbers of cats, that went away from the houses, and became wild, living among the rocks, where they find good prog, feeding on young partridges, so that they became as great a plague as the rats'.[28] Peter Mundy, noting in 1656 the 'great store' of dogs, blamed them for the difficulties his company had in locating any goats at all, the population 'supposed to bee devoured by dogges, which have mightily encreased here'.[29] Rats undoubtedly arrived as incidental introductions but they too found conditions so congenial that they multiplied at a fearsome rate to the point where they posed a major hazard to the human population as well as wreaking havoc amongst the nesting birds; they also gained a reputation for eating any crops put into the ground. By 1666 it was judged that 'there was nothing obnoxious to the amenities of life' on the island 'except vast quantities of rats, on which the Governor wages sanguinary war'.[30] Thirty years later, François Leguat commented that 'the few Inhabitants who live on this Island (in some English plantations) might enjoy an abundance of all the commodities necessary for life, were it not for the prodigious number of rats that spoil their fruit and corn',[31] and twenty years after that, in 1717, 'the rats ate up all the grain as soon as it was sown'. Only at the turn of the twentieth century was the rat population finally brought under control when the governor, an authority on the small mammals of India, gave a bounty for every one killed – starting at a penny a rat and, as they became progressively scarcer, rising to threepence.

At some unknown point a breeding population of rabbits – chiefly grey or black – was introduced: these were sufficiently numerous in the wild for Webster to mistake them (along with wild cats) as native species, although populations seem never to have reached the overwhelming proportions that occurred elsewhere.[32]

Also noteworthy, though of less direct economic significance, is the array of endemic invertebrates – described as 'startlingly diverse' – that populate the island; some of these are of exceptional size, such as the giant earwig (*Labidura herculeana*) and the giant ground beetle (*Aplothorax burchelli*), the latter named after William Burchell, who was also the first to observe the spiky yellow woodlouse (*Pseudolaureola atlantica*). In the absence of an endemic mammal

---

[28] Gosse 1990, 140.
[29] Mundy, *Relation* XXXVI, fol. 230, quoted in Mundy 1914, 330 note 2 (Ólafsson 1932, 79–80). Under Governor Patton it was required that all dogs had to wear collars identifying their owners, any caught without a collar having to be destroyed.
[30] Gosse 1990, 54. The same author (p. 167) records a report from the 1730s of 'nests, like birds' nests, built by the rats at the tops of the trees, each nest two feet across and containing six or seven young ones'; the rats overran the whole countryside around the Great Wood at this time.
[31] Leguat 1720, II, 170.
[32] Webster 1834, I, 373. The feral cat was said by Mellis (1875, 82) to exist abundantly in the wild, where, 'amongst the eggs and young of partridges and other birds it commits such havoc that sportsmen never lose an opportunity of killing it'. The rabbits remain on the island, although numbers are comparatively low.

Population and environment: early impacts

population, the insects provide valuable raw material for researchers into the evolutionary population of the island. The composition of the spider population, for example, with forty-six endemic species identified so far, has been calculated to have resulted from perhaps thirty different colonization events; this contrasts with some twenty-three colonizing stocks of beetles, taken to reflect a greater aptitude on the part of the former for trans-oceanic migration by drifting in the air as well as on the driftwood to which the latter would have been limited.[33]

Amongst the natural resources extensively harvested by the early visiting Europeans (though without detectable diminution) were those from the surrounding seas. Van Linschoten commented in 1598 on the 'great abundance of Fish, round about the Iland, that it seemeth a wonder wrought of God: for with crooked nayles, they may take as much Fish as they will, so that all the shippes doe provide themselves of Fish of all sorts in that place, which is hanged up and dried, and is of as good a taste [and savor] as any fish that ever I eate', the latter opinion being widely shared.[34] The abundance of flying-fish was particularly striking:

> About [Ascension] and the Iland of Saint Helena, unto the Equinoctiall line, there are flying Fishes, as great as Herings, which flie by great flockes together, two or three Fadome above the water, and flie in that manner at the least a quarter of a mile, untill their wings or finnes be drie ... The cause why they flie in that sort is, because they are chased by the great fishes, that eate them, and to escape from them they flie above the water, and some times into the shippes: for many of them fell into our ship, which flew too high, for when their wings are drie they must needes fall.[35]

Peter Mundy had the good fortune to hook there an enormous flying-fish: at 19 inches long, 'none in the ship ever saw a bigger'.[36] At the turn of the nineteenth century Brooke enumerated twenty-six species of fish from the surrounding seas (expanded to seventy-six by Beatson in 1816): mackerel evidently were most common, but also albacore and a number of less familiar species which he named cavalloes, coal-fish, jacks, soldiers, old-wives, and bull's eyes. The only limiting factor in harvesting them, it seems, was the small size of the fishing boats available to the islanders. At the conclusion of Captain Munden's retaking of the island from the Dutch in 1673 (Chapter 5), three boats he left behind were ordered 'to be kept in repair [in order] to permit the inhabitants to goe a fishing therein. All fish to be distributed equally amongst the inhabitants.'[37] Francis Rogers, a merchant in transit through the island in 1703, recorded that there were to be had in the roads an abundance of flying-fish and small mackerel, 'which we call horse mackerel,

---

[33] Ashmole and Ashmole 2000, 81–89.
[34] Van Linschoten 1885, 256.
[35] Ibid., 261.
[36] Mundy 1914, 331.
[37] Gosse 1990, 75.

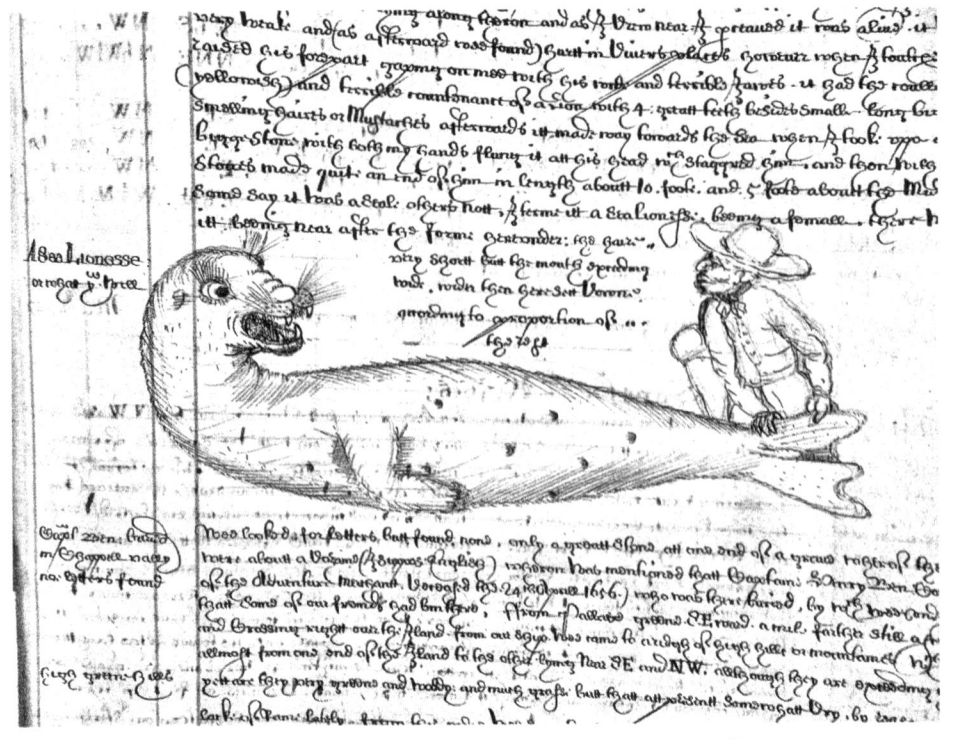

Fig. 5 Peter Mundy's encounter with a 'sealionness' near Chapel Valley, May 1656; it was 'in length about 10 foote and 5 foote about the middle'.

and take mostly in the night with a candle and lanthorn in our boat, with hook and line, in great numbers'.[38]

Webster commented in the 1830s on the numbers of sharks in the vicinity: 'The whole family of sharks are found there; the blue shark, the dog shark, the hammer-headed shark, the copper-headed shark, and the mackerel shark, all herd together, hungry for prey'.[39]

Whales – humpbacked and sperm – were seasonal visitors from July to September; grampuses and dolphins were common. Other marine mammals may have formed colonies on the island, but their vulnerability to human predation quickly rendered them scarce. Any established colonies of seals would certainly have vanished in this way. Mundy recorded killing 'a strange creature' which he identified as a female 'sealionness' (Figure 5) – now thought perhaps to have been a southern elephant seal (*Mirounga leonine*) – which was probably an isolated and sick animal; other references to sea cows (and even manatees – as in Manatee Bay) probably relate to similar species. Marine mammals contributed from time to

[38] Ibid., 120.
[39] Webster 1834, I, 376.

time to the early economy but, as mentioned earlier, their vulnerability to human predation quickly rendered them scarce.

Turtles were caught seasonally, according to Webster by taking them while asleep on the surface of the water and turning them on their backs – no mean feat since they weighed in at up to 500 lbs. Their presence on the island itself is rarely recorded, although it has been suggested that this may be due to their having favoured the remote Sandy Bay beach as their nesting site; both turtles and eggs were harvested there in small numbers until recent times.[40]

There were also shellfish and crustacea – long-legs and stumps, crabs, crawfish – and rock-oysters were common in some places.[41]

While alarm bells were already ringing concerning some of the more obviously threatened species, there remained at the time of the first settlement of St Helena a sufficiency of natural resources as to encourage the idea that it might in some respects prove able to support a human population, but the peopling of the island would only bring new pressures to bear on what would prove to be a highly precarious ecosystem.

---

[40] Ashmole and Ashmole 2000, 89–90.
[41] Gosse 1990, 75; Brooke 1808, 20–21.

4

Population and environment: asserting control

Permanent settlement of the island so profoundly changed its face that within half a century the continuing viability of maintaining a foothold there was called into question. At his arrival as governor in 1714, Isaac Pyke was moved to wonder whether the population then established might not be better decanted in its entirety to Mauritius, which had recently been abandoned by the Dutch. For his ultimate superiors, the Court of Directors of the East India Company, however – no matter their disappointment at the seeming incapacity of the island to sustain itself independently – St Helena's strategic significance ruled out such a possibility.[1] The viability of the island as a permanent settlement would continue to be questioned for decades to come.

Populating the island

Having received a charter to govern St Helena from Oliver Cromwell in 1657 (all copies of which appear to have been lost),[2] the directors of the EIC – whose homeward-bound vessels had for the past nine years been formally ordered to assemble there to form convoys for the final leg of the voyage northwards – now found it expedient to establish a permanent presence on the island.[3] Having debated the proposal on 15 December 1658, the directors decided on a show of hands 'to send 400 men with all expedition to remayne on the island, with conveniences to fortifie and begin a plantation there'. Accordingly, a fleet commanded by Captain John Dutton sailed for St Helena the following February, escorted by the *Marmaduke* man-of-war, though evidently carrying rather fewer men than originally envisaged. Dutton was

---

[1] Grove 1993, 328.
[2] In publishing a petition from the Company to Cromwell, dated only a few months later, in November 1657, William Griggs (1909, 12), editor of *Relics of the Honourable East India Company*, observes of the original charter that 'after the Restoration [it] was suppressed with such thoroughness that not only the original but all the copies have disappeared'.
[3] Preparations had been at an advanced stage for the dispatch of a fleet under Captain Dutton to colonize Pulo Run in the Spice Islands (Moluccas), when a change in the political situation there led to its cancellation; the Company seems simply to have reassigned that expedition to take possession of St Helena as an interim measure, on the successful completion of which Dutton would be carried on to Pulo Run (*Calendar of Court Minutes*, 302).

Population and environment: asserting control

to become the first governor of the island (1659–61), with authority over the whole population of 'planters' – the term accurately reflecting both the population's status as Company-sponsored migrants and the expectation that one of their immediate priorities would be to render the settlement economically self-sufficient.

With this aim in mind, Captain Bowen of the *London* was instructed to set a course via St Jago on the Cape Verde Islands in order to acquire 'all manner of plants, roots grains, and all other things necessarie' for the establishment of the plantation, as noted in the previous chapter. Once this bounty had been safely landed at St Helena (detailed instructions having been issued for their preservation on board ship), further orders from the directors declared that:

> you are espetially to have regard to the first season and opportunitie that God shall graunt unto you to proceede to planting of your provisions, but espetially your plantans and cassada, because they otherwise will be in danger to perish; and this doe in severall places of the island as you shall find convenient. And that you also proceede to set your carpenters and other artificers on worke for the framing and getting up your magazine and storehouse and other houses necessarie for your accommodation within the lynes of your fortifications.

Despite all this careful planning, the Company's ambitions to make the community self-sustaining proved over-optimistic,[4] for the early planters were supplied with repeated cargoes of foodstuffs from England, India and Africa; ships returning from Surat in India were ordered each to supply a ton of rice to the islanders when they called there. Neither did the new residents immediately flourish and multiply: when Dutton was reassigned to Pulo Run in December 1660 the majority of the civilian population opted to go with him, leaving his successor, Captain Stringer, with a mere thirty men to hold the entire island.

The gradual increase recorded thereafter was fuelled by the practice introduced under Stringer's regime (and already followed in other colonies) of making free grants of land to individual planters. The governor divided the island into 150 lots, with one plot assigned to each planter, fifteen reserved for the Company and five for Stringer himself. Philip Gosse comments on the almost feudal nature of the arrangement, for a condition attached to each grant was that the planter in question should render military service to the governor, turning out whenever the alarm announced the approach of a ship.[5] At the same time, elements of Utopia survived in the Company's stricture that planters were to 'live together in love and amity', even if backed with the sanction that failure to do so would lead to 'speedy removal'.[6]

---

[4] Although our emphasis here is on the establishment of an economically self-sustaining community, Philip Stern has observed that the Company's ambitions ranged much further, for they 'quickly came to imagine the island in global terms as … a cornerstone of a fluid and interconnected system of cities, ships, people, goods and ideas' (Stern 2007, 2).

[5] Gosse 1990, 50–51.

[6] Royle 2019, 46–47, 49. Royle observes that the yearning after Utopia expressed in the philosophical writings of Thomas Hobbes and others was given a wider social impetus

Following the temporary loss of the island to the Dutch in 1673 (Chapter 5), concerted attempts were made to boost the population. In the immediate aftermath, 110 persons – some soldiers and some civilians – were sent out, the planters each rewarded with twenty acres of land and two cows, as well as plants, seeds and entitlement to twelve months' provisions; it was assumed that by the end of this time period they would all have become self-supporting. A dispatch from the directors of the EIC, dated 8 March 1676, sounds a warning note:

> ... we find there is wanting industry and painstaking in many of the inhabitants which we will not permit to continue to be amongst you for they that will not plant should not eat – we will not supply them, rather send them home under the title of Drones.[7]

At the same time, the random collection of timber shacks that had grown up behind the fort was cleared away by order from London in favour of regular streets some 20 feet wide, 'and if there be any irregular buildings that obstruct the Evenes of the lines of the Streets or other decent usefull uniformity cause the Owners of such buildings to pull them down or alter them according to such method and rule as you shall judge most for the publique good'.[8] The principal streets were now lined with stone- or brick-built houses (estimated at twenty or thirty in number by a visitor in 1691 and at sixty to eighty in 1703 and seventy to eighty in 1717), together with a church and a market-place – the founding elements of today's Jamestown.[9] For building purposes the island's intractable volcanic stone was generally pressed into service, while limestone and Portland stone arriving as ships' ballast was scrupulously salvaged for architectural details.[10] Under Governor Roberts's regime (1708–11) means were found on the island for making bricks and tiles – potentially a boon to the residents and holding out the prospect of an enormous saving in freight for the EIC, although the bricks proved to be of poor quality and the lack of fuel to fire them rendered them uneconomic.[11]

Many buildings continued more commonly to be roofed with boards over which was spread a layer of mud; since several sources for this mud included

---

during this period by the recent regicide of Charles I, the bloodbath of the Civil War, the Great Plague and the Fire of London.

[7] Janisch 1885, 5.
[8] EIC to St Helena, 1 August 1683, British Library, India Office Records (hereafter IOR) E/3/90 fol. 93. The order continues that the cost of this exercise could not amount to much 'since we understand most of them are built but of 13 or 14 foot high with loose stones piled one upon another'.
[9] Many of these houses belonged to families with country properties, who would decamp to town whenever a fleet arrived, bringing the prospect of a market for their produce and a greatly enhanced social life.
[10] As early as 1666 the Sieur de Rennefort had noted a timber building roofed with tiles within the fort; Gosse (1990, 53–54) comments on them that they too 'must have served as ballast to some boat'.
[11] Later attempts at brick-making, although more successful, still failed to produce a viable industry (Kitching 1937, 3–4).

quantities of the 'terra Puzzolana' that proved so usefully resistant to penetration when compressed by ramming,[12] these roofs could be reasonably waterproof (though prone to harbouring rats). From the turn of the nineteenth century an alternative practice predominated – of roofing with wooden shingles, fashioned from the staves of disused barrels that hitherto, when emptied of their contents of salt meat, flour, etc., had served merely as firewood. From the late eighteenth to the mid-nineteenth century, tarred paper roofing was also common on the island as it was in Britain.[13] Prominent in the long list of dire shortcomings by which Napoleon's quarters at Longwood House were characterized was the fact that 'the roof of this hovel consists of paper, coated with pitch, which is beginning to rot, and through which the rain-water and dew penetrate'.[14] Thatch was also widely utilized:[15] Captain Bligh described most of the houses in Jamestown as being thatched in 1792, while the Longwood windmill, photographed in the mid-1800s (see Figure 9), was similarly roofed.

A priority of Governor Roberts at his appointment in 1708 was to reform the regulations under which planters held their land from the Company. Not only did he arrange for the codification of these agreements, many of which had been forgotten or ignored for years past, but he took steps to enforce them. An annual charge of one shilling, which should have been levied for every head of livestock pastured on the wasteland, had been unpaid for years until Roberts confiscated a number of cattle belonging to the widow of a previous governor, sending an unequivocal message to other landholders. Lack of fencing, which had led to such widespread damage being inflicted on the environment by unconfined herds of sheep, goats and cattle, also attracted his attention. The considerable measure of success he achieved during the four years of his governorship owed a great deal to the efforts he made to persuade the population (now standing at around 700, including the garrison) of the wisdom of his policies. It seems remarkable that in such a short period, in the words of one historian, 'Out of disorder and chaos he evolved a law-abiding, temperate and flourishing little colony'[16] – a characterization that, admittedly, would come to be seen as a little premature.

Natural disasters continued to dog the efforts of successive governors to regulate the economy of the island. In 1718 a deluge carried away soil, grass, trees

---

[12] The term was used to refer to a siliceous clay named after Pozzuoli on the slopes of Vesuvius which, when mixed with lime, water and pumice, formed a cement widely used from the Roman period onwards.
[13] For the technique see Airs 1998.
[14] *Napoleon's Appeal* 1817, 2.
[15] William Roxburgh (in Beatson 1816, 309) recorded finding '*Fimbristylis textilis* ... St Helena thatching rush; is a native of the interior of the island, and in plenty for every purpose ... A good substantial covering of this rush is said to last from 10 to 15 years, and keeps out wet effectually.'
[16] Gosse 1990, 130.

and stone walls, leaving many families ruined.[17] Droughts were more common and just as devastating in their impact: a census drawn up soon after the arrival of Governor Pyke showed that one episode had left the island with only sixty head of cattle, twenty-four pigs, three sheep and twenty-six fowls, causing the governor to decree a diet of salt meat and fish on two days a week each. Such were the pressures on seabirds' eggs during this period that the birds retreated forthwith to the most inaccessible cliffs to build their nests.[18]

Slaves and near-slaves

If life on the island could be challenging for some, there was another element of the population for whom even the uncertain life of the planter would have seemed enviable, namely the enslaved Africans and others who had formed an unwilling presence since the earliest days of British rule and whose numbers continued to expand thereafter.

At first the introductions of enslaved labour were sporadic and adventitious, but by 1684 the EIC could be found expressing the conviction that 'it is utterly impossible for any European Plantation to thrive between ye Tropicks upon any place without ye assistance and labour of Negroes'.[19] In the course of the eighteenth century, numbers of the forcibly imported enslaved population would graduate to the status of 'Free Blacks', and by the time the Slave Emancipation Act was enacted in 1834 many of those living on the island were third- or fourth-generation descendants of earlier migrants, while not a few had come to share common blood with the European population.

The shadow of slavery appears to have been cast over the island with the arrival of its very earliest European inhabitant, for it seems that Fernão Lopez (Chapter 2) was accompanied by a Javanese slave.[20] According to one source, at some time before 1557,

> two kaffirs from Mozambique and a Javanese man and two women slaves swam ashore from a ship and hid themselves in the woods until the vessel had sailed ... For a long time the blacks defied all attempts to capture or destroy them.[21]

Arriving on the island in the same year, the Patriarch of Abyssinia (Chapter 2) found that 'certain fugitive slaves' – perhaps this same group – who had earlier murdered a chaplain, began to seduce his own slaves so that they would no longer work for him. Given the density of the forest cover, the inaccessible nature of parts of the island and the intermittent nature of Portuguese visits, it is entirely possible that some or all of these persons could have survived until the time of Thomas

---

[17] Ibid., 134.
[18] Ibid., 135.
[19] IOR E/3/85, ff.94–95v, quoted by Royle 2007, 84.
[20] Birch 1880, xxxviii. Brooke (1808, 37) has Fernão Lopez being sent into exile and landing 'at his own request on St Helena with a few negro slaves' – seemingly an unlikely indulgence.
[21] Gosse 1990, 11, without source.

Cavendish, who noted in 1588: 'We found in the houses at our comming 3 slaves which were Negros, & one which was borne in the yland of Java.'[22]

When the EIC gained control of the island, deliberate attempts began to be made to expand this element of the population. The instructions given to Captain Bowen on the initial voyage of settlement included a charge to acquire at St Jago 'five or six blacks or negroes, able men and women', provided they were to be had for no more than 40 dollars each.[23] The following year the captain of the *Truro* was directed to the coast of Guinea in order to acquire a further 'ten lusty blacks, men and women';[24] three years later Captain Swanley was directed to embark a further '12 lusty young Negroes, the major part women'. Although the term 'slave' is absent from the Company records at this time, and despite a record of instructions being issued in relation to those embarking at Fort Cormantine (site of an English factory since 1638, lying in present-day Ghana) that they should be limited to 'such as will voluntarily and without compulsion sail in ye ship',[25] it might seem remarkable that men and women who had already been introduced to the brutality of the slave trade would willingly consign themselves to the sea in this manner. Nonetheless, the EIC directors of the 1670s were specific in their instructions that those who arrived in this manner were to be classed as 'black servants' and not as slaves, while compared to later decades there was a willingness under the 'levelling' constitution of the early years of British rule to treat the black population with a degree of consideration: children were to receive a basic education and adults who were received into the Christian church and had worked for seven years would gain their liberty and equal status with the white population.[26] On the other hand, the instruction to overseers not to be 'too cruel' to those working under their supervision, while outwardly liberal in intent, carries chilling implications.

Following an invasion by the Dutch in 1672 and the subsequent reoccupation of the island by the British (Chapter 5),[27] the scale of immigration as well as the tenor of society changed significantly. In the course of the following decade the black population came to outnumber the white, giving rise to anxieties that the Europeans

---

[22] Hakluyt 1903–05, XI, 346. According to Gosse (1990, 11) the earlier handful of enslaved immigrants had in the meantime multiplied to twenty.
[23] Foster 1919, 285.
[24] Gosse 1990, 50.
[25] Royle 2007, 40, 85.
[26] Michael Bennett (2021) gives a detailed account of how the Company came to see the benefits of such a policy in its dealings with other societies – perhaps especially in the Mughal empire – as very much in its own (primarily mercantile) interests.
[27] One of the heroes of the reconquest was 'Black Oliver', a slave who had been evacuated with his owner and sold by him in Brazil before being recovered as a crew member for Captain Munden's voyage back to St Helena; there he acted as a guide to Captain Keigwin's landing party, disembarked successfully at Prosperous Bay (see Chapter 4). He was rewarded with a grant of land and livestock, but was later shot while taking part in a riot outside the fort in 1684.

might be overwhelmed in the event of an uprising,[28] though in truth more danger to the stability of society would always come from malcontents among the white civilian population and the military. Nonetheless, restrictions were imposed from time to time on the numbers of slaves permitted on the island, although immigration never stopped entirely. The Royal African Company collaborated in the supply of those from West Africa, but by the 1680s others were being shipped from the island of Johanna (Comoro Islands) and from Madagascar. Approval was given in 1681 for the refreshment at the island of so-called 'Madagascar ships' – slavers bound for the Americas – the services rendered to be paid for in slaves, but soon it was found that Company ambitions to establish plantations and industries that might not only supply the needs of the island but could also form the basis of a profitable export market could not be met in such a piecemeal manner. In 1684 Captain Knox was sent to Madagascar to buy slaves with a view specifically to introducing to St Helena plantations for sugar, tobacco and indigo.[29] At the same time a West Indian supervisor was established there and three years later fifteen slaves who had become proficient in sugar cultivation in the West Indies were reassigned to St Helena. In the same year, ten slaves were brought from India with a view to establishing mountain rice as a staple crop on the island.[30] In order to implement his ambitious programme of planting and enclosure, Governor Roberts estimated in 1708 that the slave population should be increased by a further 200 souls – a consequent expansion of the acreage under cultivation being required to supply the increasing number of yams that formed their almost invariable diet.

By the mid-eighteenth century Madagascar had become firmly established as the favoured source of slaves, the population there having proved by experience to be less truculent and more amenable to sustained work.[31] We may easily appreciate that enforced transplantation would have induced high levels of anguish and despair, but further fuel was added by the fact that the European population

[28] Rumours of imminent revolt surfaced from time to time – usually without foundation – prompting the most savage reprisals.
[29] Earlier, in 1666, Sir George Oxenden had sent indigo seeds from Surat to St Helena, but warned the Court of Directors in London that 'we cannot possibly gett a person here to direct them in the sowing & making indigo: those that plant it here being all natives & have families, who will not be persuaded to leave them & their Relations on any consideration'; Oxenden recommends that they 'might endeavear the procuring of a Person or two out of those plantations of Barbados etc.' (Sir George Oxenden and Council in Surat to EIC in London, 1 January 1666, quoted in Winterbottom 2010, note 773).
[30] Royle 2007, 85. Twenty years earlier the Court of Directors requested their representatives at Fort St George to 'procure young Gentues or Aracans [i.e. Hindus and Burmese] and their wives to be sent as our servants to remain on our Island of St Helena Wee being very desirous to make trial of them, supposing that they may bee more usefull and ingenious than those people which come from Guinea'. These specifically were not acquired as slaves, however, and were to be limited to 'such as will willingly embrace our service' (Winterbottom 2010, note 850).
[31] For a time, every vessel touching at Madagascar on the voyage to St Helena was obliged to contribute one slave to the island (Brooke 1808, 80).

had come to believe that only the lash (or the threat of it) would induce enslaved Africans to any kind of industry. Flogging was comparatively commonplace; in the case of one offender, 'after forty stripes administered on his naked body [he was] to have a pair of iron pothooks riveted around his neck until further orders'.[32] Hands were cut off for more than one infraction, and sealing wax was dropped on the skin for others, while offences considered of a serious nature brought punishments of the most barbarous kind: in 1695 one putative rebel was hanged in chains on Ladder Hill and left there to starve to death, while two of his co-conspirators were hung, drawn and quartered. Others were burnt at the stake; in 1685, while two slaves were burnt alive, a woman accused of being an accessory was bound to a post close to the fire, afterwards being given thirty lashes and returned to prison to be kept in irons until she could be deported elsewhere, 'but whilst she stays, to receive 30 lashes on her naked body every Saturday afternoon'.[33] Women were whipped naked through the street on more than one occasion; Mellis records one Sarah Marshall, given 'one-and-thirty lashes on her naked body at the flagstaffe for scandalizing Captain Bendall'.[34]

A glimmer of compassion (or perhaps self-interest) might be detected in a pronouncement of Governor Pyke at the trial of one of the Company's slaves, caught trying to break into the storehouse:

> we find by experience among the planters [that] most of them [are] very severe to their slaves, but we can't perceive that it does them any good, but rather makes them worse, always solemn, often desperate, and in their despair they sometimes hang or drown themselves, or run away.

Whichever sentiment moved the governor on that occasion, it seems to have made no long-term impact. On several occasions slaves (and occasionally deserters from the garrison) made off in hopelessly inadequate open boats, without adequate provisions or equipment: almost all were never heard of again.[35]

Joseph Banks was one of those deeply offended by the oppressive conduct of the islanders and by the barbarity with which discipline was maintained. He commented:

> Their slaves indeed are very numerous: they have them from most parts of the World, but they appear to me a miserable race worn out almost with the severity

---

[32] Mellis 1875, 8.
[33] Gosse 1990, 94.
[34] Mellis 1875, 8. As on other occasions, treatment of the military personnel could be similarly brutal, often involving the painful sentence of 'riding the wooden horse': swearing and incivility earned one William Melling two hours on the horse 'with a bag of shott at each heele', while for 'slighting the Government and malitiously revenging himself' Richard Honeywood rode the horse for half an hour 'with two muskets at each heele' (ibid.).
[35] In 1772 a slave named John Fortune arrived back on the island, the boat in which he was fishing having been commandeered two years earlier by seven soldiers and sailed to Brazil with him as crew. (Prior (1819, 89) mentions that they survived the passage only by killing and eating one of their number.) Fortune made his way back to the island via London, where he had surrendered himself to the East India Company.

Fig. 6 Good conduct medal, obverse and reverse, awarded to an enslaved person, Samuel Caesar, 1824.

of the punishments of which they frequently complaind. I am sorry to say that it appeard to me that far more frequent and wanton Cruelty were exercisd by my country men over these unfortunate people than even their neighbours the Dutch, fam'd for inhumanity, are guilty of.[36]

At the time of his visit on Cook's second voyage in 1775, however, George Forster found comparatively few complaints from the slaves he questioned, beyond the fact that they were only sparingly supplied with food and at some seasons had to subsist on salt provisions: tellingly, Forster thought the situation of the soldiers in the garrison was even worse, 'they being confined to constant salt-diet, of which the East India company ... allows very scanty portions'.[37] Like the soldiers, those slaves completing long service and displaying good conduct might be awarded a medal (Figure 6) – surely the most condescending of rewards for a lifetime of enforced servitude.[38]

Most of these unwilling immigrants became the property of individual planters (Figure 7), while some remained the property of the Company;[39] others

[36] Banks 1962, II, 267. In the 1720s slave owners had successfully petitioned the governor for the right to inflict corporal punishment on runaway slaves at their own discretion and without recourse to the courts.
[37] Forster 2000, II, 664.
[38] These medals were introduced under Governor Patton, who was generally liberal in his attitude to slavery, but they were of very limited value: Gosse (1990, 242) mentions that 'if a slave should win one of these coveted prizes in three successive years, he did not, as might have been expected, receive his freedom but only "have his merits completely established"'.
[39] Whatever the poor view of African labour held by the planters, an early observation from Henry Gargen in the earlier 'more benign' period of the 1660s is revealing. Gargen for his part held a low opinion of the European planters that would long be shared

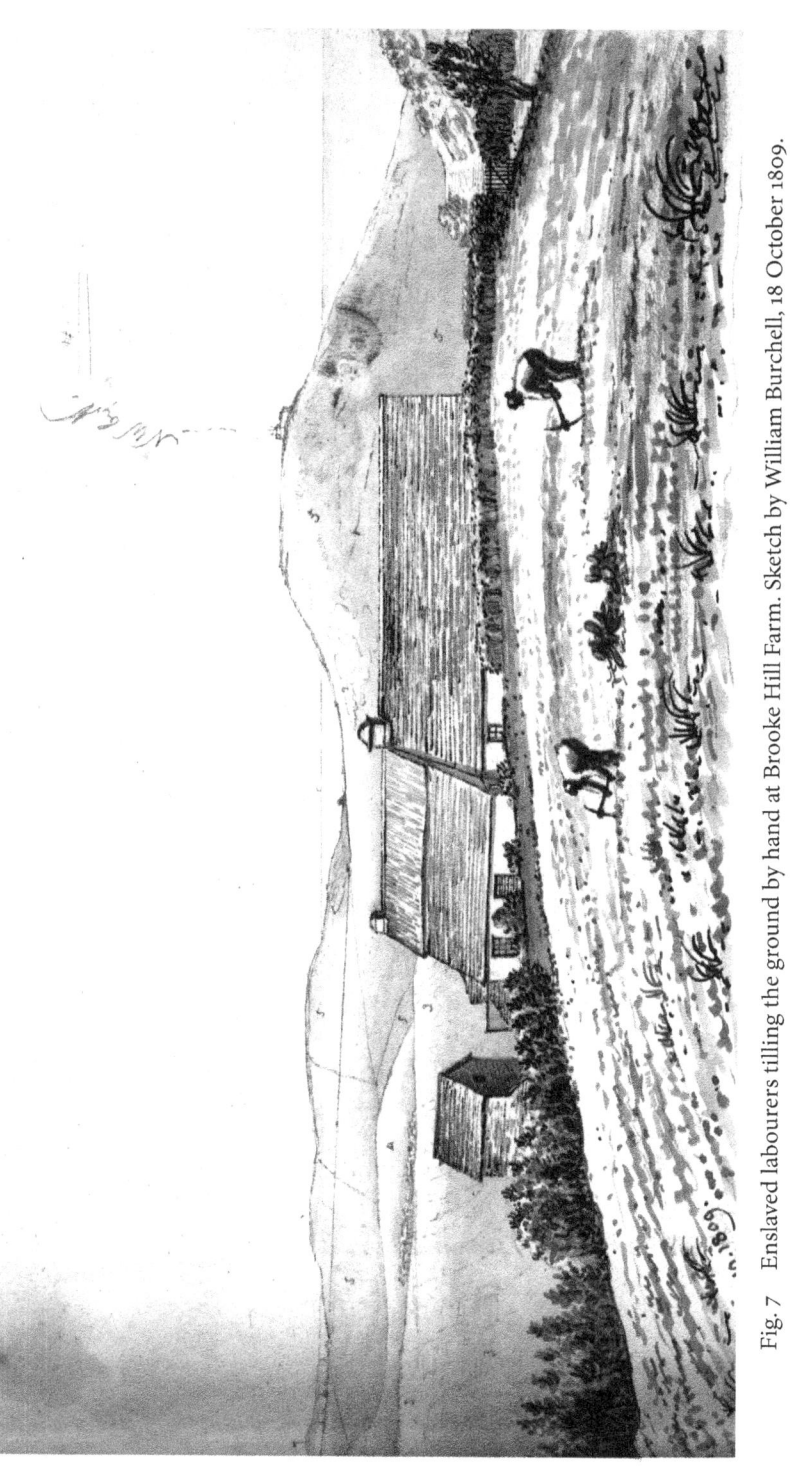

Fig. 7 Enslaved labourers tilling the ground by hand at Brooke Hill Farm. Sketch by William Burchell, 18 October 1809.

entered domestic service or a specific trade. Soldiers were at times tasked with teaching them new skills, although until the last days of slavery on the island most would remain without a recognized profession. Colin Fox's account of the latter days of slavery on the island, *A Bitter Draught* (2017), includes not only detailed analyses but also extensive tables recording the names and 'qualitys' of those involved and also their employment. The men worked as stonemasons, carpenters and lime kiln workers involved in maintaining the fortifications and watercourses, as boatmen, gardeners and house servants, as stockmen, a great many as undifferentiated labourers, and only a handful in trades – tailors, smiths, sawyers, carpenters, butchers; the women as washerwomen, dairymaids, a seamstress and a midwife. Some are judged very competent, others indifferent, sickly, 'much poxed' or 'good for nothing'.[40]

The end of slavery in St Helena came about incrementally, prompted, no doubt, by the increasing effectiveness of the anti-slavery movement in Britain. In 1792, as part of a package of reforms that brought about improvements to the conditions of those already on the island – limiting the severity of punishments that owners were allowed to mete out to slaves while extending the powers of the magistracy to prosecute those who exceed their 'rights' – Governor Brooke succeeded in bringing an end to the further importation of slaves to the island.[41] By 1807 the trading of slaves was banned throughout the British Empire, though again without materially affecting the status of those already enslaved. Then in 1818 Governor Sir Hudson Lowe persuaded the slave owners to agree that (following a similar arrangement recently introduced in Ceylon) from Christmas of that year all children born to enslaved parents on the island would have their freedom acknowledged. Six years later, when a locally inspired recommendation was made to Governor Alexander Walker that a poll tax should be levied on all free blacks, he pointed out not only the impossibility of introducing new legislation based on colour but also the fact that there was by that time a population amounting to some hundreds of individuals amongst whom it would be impossible to decide if they were black or white.

While the directors of the Company could conceivably have been swept along by the reformist movement, there was a growing discomfort amongst them at finding themselves cast in the role of slave owners. During this period of flux, certain slaves gained the right to purchase their freedom with money that could be borrowed from the East India Company – an arrangement unique to St Helena, it would appear – while others continued to be bought and sold in the market (Figure 8). In 1832 the Company abolished slavery on the island altogether and purchased the freedom of 614 enslaved individuals, compensating their owners a

---

by the EIC directors in London; he commented that the slaves who farmed less than one-third of a plantation 'hath three times as many beanes. And one of their beanes is worth five of theirs, both for planting and eating' (Royle 2007, 86).

[40] Fox 2017, appendices 8–15.
[41] Summarized in ibid., table 1.

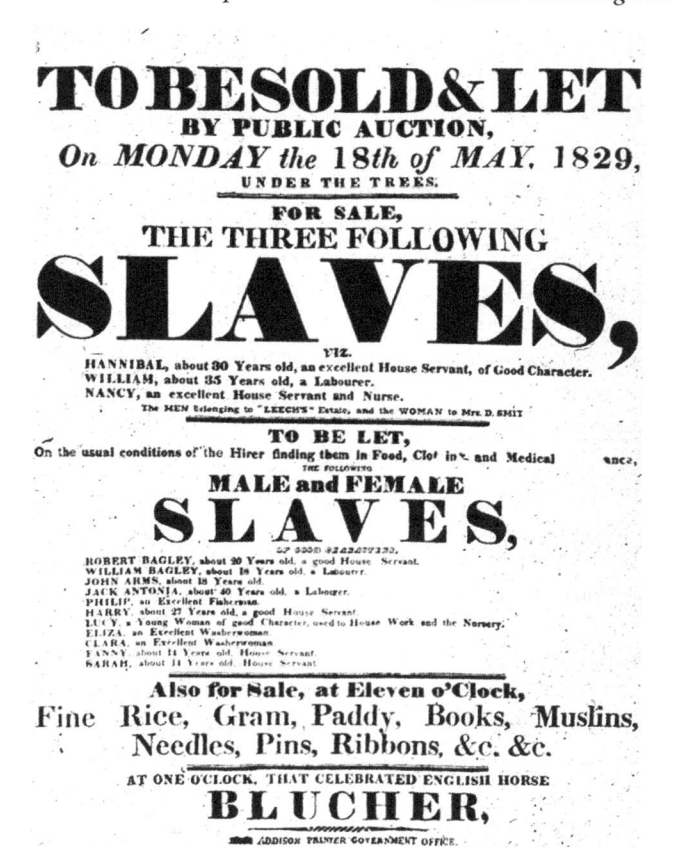

Population and environment: asserting control

TO BE SOLD & LET
BY PUBLIC AUCTION,
On MONDAY the 18th of MAY, 1829,
UNDER THE TREES.

FOR SALE,
THE THREE FOLLOWING
SLAVES,
VIZ.
HANNIBAL, about 30 Years old, an excellent House Servant, of Good Character.
WILLIAM, about 35 Years old, a Labourer.
NANCY, an excellent House Servant and Nurse.
The MEN belonging to "LEECH'S" Estate, and the WOMAN to Mrs. D. SMITH

TO BE LET,
On the usual conditions of the Hirer finding them in Food, Clot in and Medical ance,
THE FOLLOWING
MALE and FEMALE
SLAVES,

ROBERT BAGLEY, about 20 Years old, a good House Servant.
WILLIAM BAGLEY, about 18 Years old, a Labourer.
JOHN ARMS, about 18 Years old.
JACK ANTONIA, about 40 Years old, a Labourer.
PHILIP, an Excellent Fisherman.
HARRY, about 27 Years old, a good House Servant.
LUCY, a Young Woman of good Character, used to House Work and the Nursery.
ELIZA, an Excellent Washerwoman.
CLARA, an Excellent Washerwoman.
FANNY, about 14 Years old, House Servant.
SARAH, about 14 Years old, House Servant.

Also for Sale, at Eleven o'Clock,
Fine Rice, Gram, Paddy, Books, Muslins, Needles, Pins, Ribbons, &c. &c.

AT ONE O'CLOCK, THAT CELEBRATED ENGLISH HORSE
BLUCHER,
ADDISON PRINTER GOVERNMENT OFFICE.

Fig. 8 Poster for the sale and letting of slaves on St Helena, 18 May 1829. Aged from 14 to 40, the men, women and children are given more perfunctory treatment than the 'celebrated English horse Blucher'.

total of £28,062 17s. Another eight years would pass before the Treasury decided to write off those loans for manumission that remained unredeemed – amounting to some 90 per cent of the total sum lent.

In 1833 control of the island reverted to the Crown, removing the last vestiges of Company rule. While endorsing the new regime, Charles Darwin mused on the continuing problems of sustainability faced by the newly liberated population:[42]

> The public expenses, if one forgets its character as a prison, seems out of all proportion to the extent or value of the Island. So little level or useful land is there, that it seems surprising how so many people (about 5,000) can subsist. The lower orders, or the emancipated slaves, are, I believe, extremely poor; they complain of want of work; a fact which is also shewn by the cheap labour. From the reduction in number of public servants owing to the island being given up by the East India Company & consequent emigration of many of the richer people, the poverty probably will increase … the fine times, as my old guide called them,

[42] Darwin 1933, 411.

when 'Bony' was here can never again return. Now that the people are blessed with freedom, a right which I believe they fully value, it seems probable that their numbers will quickly increase: if so, what is to become of the little state of St Helena?

Darwin's unease proved well founded, for in the years of retrenchment that followed both the end of Company rule and the withdrawal of much of the colonial infrastructure that had provided a framework for prosperity, the former enslaved element of the population suffered greater hardship than any other.[43]

Reprehensible as any form of slavery must clearly be, the EIC had occasionally gone further than most in attempting to enslave prisoners of war who – even in pre-Geneva Convention terms – might reasonably have expected more civilized treatment. So it was that in 1757 ten men had been brought from Malabar[44] who reportedly had served as officers of the army of the king of Travancore before falling into Company hands. One of them died on the voyage from India while the others found themselves dressed in slaves' clothing and sent to work in the plantations. The experience proved too much for these cultured individuals: four of them hanged themselves soon after arrival and the others threatened to follow suit.

As the EIC extended its sphere of influence on the Indian sub-continent and expanded the number of its officers proportionately, St Helena found itself the recipient of further Indians casually abandoned by their British masters in the course of their journey home following service in India.[45] Although born free, numbers of these now found themselves sold into slavery on the island, though several successfully challenged the unwelcome status thrust upon them. In 1794 the Company formally outlawed the 'unlawful and unjust' sale on St Helena of the free inhabitants of Bengal and the other provinces of India, bringing this

---

[43] Commenting on the unhappy consequences of the abrupt change of regime, Mellis (1875, 29) comments: 'So hard did the Company's treatment of their servants press upon many of them, that twenty years after this event officers of high rank might still be seen digging the soil side by side with their own negro servant in the struggle to support their families.'

[44] Indian slaves appear to have reached the island as early as 1673, when Fernandez Navarette, a Franciscan friar returning to Spain after nearly thirty years' service in the East, met there 'some blacks from Madraspatam, for whom I was concern'd, because they had bin Catholicks at home and were Hereticks there'. He goes on to say that 'there were also two Frenchmen in the same way'; it might be thought that this could refer to no more than their faith, but at least one Portuguese renegade landed at the island in 1719 is said to have been condemned to work as a slave there for the remainder of his life. See Gosse 1990, 68–69, 149.

[45] Although remarkably callous, the practice should be seen in the context of the abandonment of 'hundreds, perhaps thousands' of *ayahs* in particular by returning British families, once they had reached England. See https://www.theguardian.com/culture/2020/mar/01/one-way-passage-from-india-hackney-museum-colonial-ayahs-london.

particular chapter to a close.⁴⁶ A separate element of the Indian population of the island consisted of a group known solely from a passing reference in the journal of Johann Reinhold Forster in 1775:

> The East-India-Company make use of this Island for a place of banishment of such people in India as oppose their measures in any way. There are now several Bramines, who were suspected to hinder somewhere in India the Success of Trade, & they were seized & transported hither, where they have in the middle of the Country a House & some Land & Gardens, with Servants to serve them, & all the Provisions & Necessaries allowed them by the Company.⁴⁷

One group that enjoyed (comparatively speaking) a more enlightened reception on the island were the indentured Chinese tradesmen who arrived for a limited period in the earliest decades of the nineteenth century to make up for the shortfall formerly supplied by skilled slaves. Their story has been researched by Po-ching Yu, who quotes sources that bracket most of the era during which the Chinese arrived. The initiative to seek Chinese labour had been taken by Governor Patton and proved immediately beneficial. In a letter from the council of St Helena to the Canton factory, dated 18 January 1811, the council acknowledged that:

> The satisfactory manner in which you have complied with our application regarding the Chinese deserves our thanks, we find these people so extremely useful that we now request you will send us 150 more. Of whom it would be desirable that a large proportion should be husbandmen and gardeners and at least six stone cutters and 12 stone masons, carpenters and blacksmiths.

In the following years further requests were sent to Canton for carpenters and stonemasons, the numbers requested ranging from some tens to hundreds of people at a time, up to the point where in 1820 Canton asked that no more such requests should be sent. At one point the Chinese population was said to stand at almost 650.⁴⁸ Most of these immigrants were considered beneficial to the economy, although some were evidently unskilled peasants rather than tradesmen and their productivity was considered poor. Chinese seamen were also abandoned from time to time on the island, but most of these appear to have been deliberately left behind on the grounds that they had proved lazy and unproductive. Ultimately the EIC concluded that the £12,000 a year it cost them to maintain this workforce was not money well spent: the practice of shipping Chinese to St Helena was abandoned and, while some remained to be integrated into the population, their

---

⁴⁶ Major 2012, 104–07.
⁴⁷ Forster 1982, IV, 574. Gosse (1990, 172) mentions also a group of natives of the Maldives whose boat had been driven out to sea in a storm; although rescued by the East Indiaman *Drake*, they too found themselves enslaved on their disembarkation at St Helena.
⁴⁸ Gosse 1990, 246.

numbers dwindled over the course of the 1820s.[49] In 1875 Mellis could record that 'the only records of their time exist in the Chinese cemetery, at a spot called New Ground, and an extremely picturesque little Jos house at Black Square; but both of these are fast falling into decay'.[50]

Feeding the population

An original condition of land grants was that every 10 acres apportioned should support one cow – a requirement that was initially enforced; those who failed to comply were deprived of their lands as unproductive 'drones'. Early planters had been allowed no more than 40 acres each, but over the years many property transfers had resulted in farms several times that size. From the time of the governorship of Captain Pyke the Company sought to reimpose these limits, to ensure that farms were individually occupied, and conversely to prevent them being broken up by inheritance by more than one person. The stock of cattle increased in number to about 400 head under Governor Byfield. These were always at risk from the droughts that visited the island fairly regularly, as well as from disease: in the early 1760s an outbreak of distemper wiped out nearly the entire population of cattle. From time to time, attempts were made to enforce enclosure of the livestock (particularly goats and pigs) kept by the planters, but the temptation to allow them to wander until they were needed (much to the detriment of the vegetation, as mentioned earlier) proved hard to resist.

Chickens, ducks, geese and turkeys were raised, proving popular items with the crews of visiting ships to whom they were sold – or often bartered for textiles and items of clothing (always in short supply on the island). They too were prone to fowl pest: in the mid-1700s an epidemic wiped out 1,000 turkeys and most of the fowls on the island.[51]

Of the root crops imported, yams proved the most enduringly successful. Their cultivation was intimately connected with the operation of a slave-based economy. When Governor Roberts and his council resolved in 1708 that the island needed an additional 200 slaves (to augment the seventy-six already on the island) their first step was to explore what grounds might be available for cultivation of the yams needed to support them: they were forced to conclude that without the introduction of extensive artificially irrigated plots, the island could not produce an adequate supply. An area of hillside between Friar's and Breakneck Valley was selected and the work was begun, with some success, but for reasons unknown the site was later abandoned and allowed to revert to its former state.

The particular yams cultivated on St Helena attracted little approval from Joseph Banks:

---

[49] Yu 2015, 298, 302.
[50] Mellis 1875, 18.
[51] Gosse 1990, 187.

Population and environment: asserting control

Yams, the same as are called Cocos in the West Indies, is what they chiefly depend upon to supply their numerous slaves with provision; these however are not cultivated in half the perfect[i]on that I have seen in the South Sea Islands, nor have they like the Indians several sorts many of which are very palatable, but are confind to only one and that one of the Worst.[52]

By the 1690s the lengthy cooking times necessary for yams had already led to a firewood shortage so acute that some slaves were found to be spending half their working hours fetching firewood from distant spots.[53] Today, although cultivation has long since been abandoned, the island population still refer to themselves as 'yam-stocks'; the plants themselves can still be found growing wild in many of the remoter valleys.

Cereals proved less well adapted to the island, although 'Indian corn' was grown successfully by the earliest generations of settlers (subject only to the depredations of rats). In the 1740s Governor Dunbar planted experimental crops of oats, barley and wheat at Longwood: the initial results proved sufficiently promising for a barn to be erected for the harvest, but subsequent seasons saw the yield decline to the extent that the trial was abandoned.[54] Although several of the island's governors showed themselves to be in tune with the movement for agricultural improvement that swept Britain in the eighteenth century, they found themselves constrained by the island's terrain: cereal cultivation never achieved the importance to the islanders of root crops and culinary vegetables. Under Governor Walker a threshing mill was imported from England and installed at New Longwood; we hear nothing of a grinding mill at this time, but a windmill was certainly established there within the following two decades, when Thomas Deason, who farmed the land, had one built in the mid-1850s. Given the steadiness of the prevailing wind at St Helena, it evidently was felt unnecessary for it to have the capacity to rotate on its axis; instead the sails were simply mounted on a fixed, gabled extension on the roof of a thatched, brick-built barn (Figure 9).[55]

Potatoes, which potentially gave three or four crops (often two in a single year) to one of yams, came to be more widely cultivated. Early crops were described as 'mostly of the red kind', although they were reduced in number with the

---

[52] Banks 1962, II, 267. Brooke (1808, 195) refers to the prevalent species as the wet yam; he also mentions the former presence of a mountain yam, of which none survived in his day, its loss 'little to be regretted' on account of its unwholesomeness.

[53] The distilling of arak, drunk in enormous quantities by both the garrison and the civilian population, was carried out on such a scale that it too contributed to the fuel shortage. Arak was considered a necessary antidote to a generally unwholesome diet.

[54] Brooke (1808, 235) conjectures that drought or inappropriate soil may have been the cause of failure. Banks recorded the unsuccessful experiment with barley in 1771, noting that it had never been repeated in the intervening years (Banks 1962, II, 267).

[55] Information from http://sainthelenaisland.info/lostbuildings.htm. The structure was sufficiently novel for a model of it to have been displayed in the India Museum in London (MacGregor, 2023, 163 note 2).

Fig. 9 Windmill at Longwood, seemingly that built by Thomas Deason c.1858. The comparatively constant windspeed and direction at St Helena obviated the need for the sails to trimmed.

prohibition of arak distilleries. With the revival of interest in them as a food crop, 'Irish potatoes' were planted instead.[56] Potatoes had the additional advantage over yams that they proved more saleable to visiting ships: by 1802 Lord Valentia reported that an acre of land might be expected to yield 400 bushels of potatoes which could be sold for 8 shillings a bushel to passing ships.

Cassava was also introduced to the island from South America, not merely for variety but because it was said to be resistant to spoiling by flies and worms. Horsebeans and peas were sent to the island in the late 1600s to be sold to planters as food for slaves.[57]

Production was always somewhat precarious, and indeed there may never have been a means by which St Helena could fulfil the expectations placed on it by the EIC, on the one hand as an efficient victualling post for ships homeward bound, often with fresh supplies at such a low ebb that crews were at constant risk of illness and disease, and on the other as a self-sustaining colony. The islanders themselves, starved of outside company for months on end, were reputedly only too glad to abandon work on their smallholdings in order to gather in Jamestown whenever a convoy came in, as they had done in Dampier's day:

> when Ships arrive, they all flock to the Town, where they live all the time that the Ships lie here; for then there is their Fair, or Market, to buy such Necessaries as they want, and to sell of the Product of their Plantations.[58]

The custom led an early nineteenth-century visitor, Lieutenant James Prior, to form the impression that the populace had by his day all but abandoned agriculture:

> There are no farmers in the island by profession. The majority of the people, being shop-keepers, live in town; and the land being subdivided between them into small tracts, they occasionally resort thither for amusement, having neither time nor inclination to attend to the soil.[59]

More profit was to be made by householders in opening their premises in Jamestown to provide lodgings for the transient visitors, eager to spend time on shore to relieve the tedium of their voyage; they would even be entertained with 'plays, dances, and concerts'.[60]

A plateau seems to have been reached in the ability of the planters to produce more, although to outside eyes it appeared that levels of production stood far below their potential. When he called at the island in 1771, Joseph Banks added his voice to those of the critics:

---

[56] Brooke 1808, 212–13.
[57] The Company also ordered that dried fish should be kept in reserve 'for ye supply of our negroes in case of failure of their provisions by accident or dearths or ill weather'.
[58] Dampier 1937, 364.
[59] Prior 1819, 85.
[60] Duncan 1805, 204.

> The White inhabitants ... live intirely by supplying such ships as touch at the Place with refreshments, of which however to their Shame be it spoken they appear to have by no means a supply equal to the extent as well as the fertility of their soil, as well as the fortunate situation of their Island seem to promise. Situate in a degree between temperate and warm their Soil might produce most if not all the vegetables of Europe together with the fruits of the Indies, Yet both are almost totally neglected. Cabbages indeed and garden stuff in general is very good, but so far from being in plenty so as to supply the ships who touch here a scanty allowance only of them are to be got, chiefly by favour from the greater people who totally monopolize every article produced by the Island, excepting only beef and mutton which the Company keep in their own hands; and tho there is a market house in the town yet nothing is sold publickly, nor could either of the three Kings ships that were there get greens for their Tables except only Cap^{tn} Elliot the Commanding Officer who was furnishd by order of the Governor out of his own garden.[61]

Banks also criticized 'the Custom of the Indias Captains, who always make very hansome presents to the families where they are entertaind besides paying any extravagant prizes for the few refreshments they get', a practice he thought had 'inspird the People with a degree of Lazyness', although in fairness he deemed these shortcomings to be national rather than local in character.

It has been mentioned that by the end of the seventeenth century a considerable range of plants and livestock had been added to the native flora and fauna in a series of piecemeal initiatives that sought to add to the resources necessary to maintain a stable population. In the course of the 1700s these continued in an increasingly purposeful manner, though all too often systematic initiatives were thwarted by the exigencies of island life, leaving the Court of Directors in London despairing that their ambitions to make the island self-sufficient would ever be met. In truth, the policies of the directors contributed directly to many of these failures, to the extent that the Company's own governors found themselves at times trying to juggle conflicting agendas demanded by London and imposed by the realities of life on the island.

In some ways the island's population as a whole seems to have been ill-suited to farming life and to have benefited little from advances being made in contemporary Britain. Agriculture, if we are to believe the evidence of Banks, was practised at a fairly rudimentary level. He observed that:

> All kinds of Labour is here performed by Man, indeed he is the only animal that works except a few Saddle Horses nor has he the least assistance of art to enable him to perform his task. Supposing the Roads to be too steep and narrow for Carts, an objection which lies against only one part of the Island, yet the simple

---

[61] Banks 1962, II, 265–66.

contrivance of Wheelbarrows would Doub[t]less be far preferable to carrying burthens upon the head, and yet even that expedient was never tried.[62]

Banks might be thought to have been over-dismissive here, as was the publisher of Captain Cook's account of the *Endeavour* voyage, who inserted on his own account a similar statement as to the lack of wheelbarrows: as a result, when Cook returned to the island four years later on the *Resolution*, George Forster recorded that 'There are many wheel-barrows and several carts on this island, some of which seemed to be very studiously placed before captain Cook's lodgings every day.'[63]

When the reforming Colonel Alexander Beatson arrived as governor in 1808, however, his assessment of 'the arts of agriculture' as practised on the island largely conformed to that of Banks.[64] Land was cultivated essentially without the aid of animals. Beatson's early attempts at reform[65] met with resistance from the labouring population, who clung to the view that 'their forefathers had done very well with the hoe, and they saw no use or necessity for such new things'. Undaunted, Beatson went on to introduce an improved system of husbandry under the direction of an experienced Norfolk farmer; ploughs were shipped to the island, initially to be loaned freely with their teams to the farmers, whose attitudes changed in time from regarding them as 'mere foolishness' to broad acknowledgement of their ease of use and the greater abundance of the crops produced with their aid, so that 'many of the most respectable soon afterwards followed the example'.[66] To promote this

[62] Ibid., II, 267. George Forster (2000, II, 664) recorded in 1775 that 'horses at St Helena are imported chiefly from the Cape of Good Hope, and a few are now bred on the island; they are small, but travel well in this country'. An Arab thoroughbred, bought at enormous expense by Governor Brooke in the 1790s with a view to improving the island's stock, seems to have made no recorded impact. As late as 1802, George Annesley, Viscount Valentia was amused to be called upon in Jamestown by 'the fair daughters of the Governor … drawn in a light carriage by oxen, the only animals adapted to ascend and descend Ladder Hill' (Annesley 1809, 14), while as late as the 1850s, Miss Polly Mason, 'one of St Helena's chief landowners', spurned the carriage and was given to riding around on the back of an ox (McCracken 2022, 45). Donkeys provided the principal means of agricultural transport, although by the later 1800s some 225 horses were kept on the island and were considered 'a necessity, and not a luxury' (Mellis 1875, 83).

[63] Forster 2000, II, 662n.

[64] McCracken's characterization of Beatson as a 'colonial Coke of Norfolk' (McCracken 2022, 114) seems entirely apposite.

[65] Beatson (1816, 11–13), who experimented with methods of planting and manuring at Plantation House, in order to determine the most effective approach (he became an enthusiastic advocate of seafowl guano), corresponded with Sir John Sinclair, a leading light in the agricultural reform movement.

[66] Beatson (1816, lxviii, n.) records that Governor Dunbar (1743–47) had experimented with the plough and had introduced the cultivation of oats, barley and wheat, 'but his successor had no taste for improvements, and those promising beginnings were totally forgotten'. At the time of his visit, Prior (1819, 86) records that, 'It is not long since there was not a plough in the island, and now, owing to [Beatson's] exertions, there are

movement the Company's farm at Longwood was utilized by Beatson to provide object-lessons for the farming community:[67] a number of 10- to 20-acre fields there were broken up into plots, ploughed by his newly recruited agriculturalists and sown with cereals or potatoes to provide crops that far exceeded those gained with the aid of the traditional mattock (see Figure 7). This initiative was consolidated by Beatson's successor as governor, Mark Wilks, who expanded the cultivated area at Longwood by a further 30 acres as well as forming a 36-acre plantation for trees. Further plantations and nurseries flourished at this time at the governor's residence at Plantation House (Plate 10), but a decade later Governor Walker was still struggling to motivate the island's farmers: he introduced agricultural shows and ploughing matches to the island, with prizes for the best produce; he founded an Agricultural and Horticultural Society and established a daily market in Jamestown.

Beatson himself published a series of short 'agricultural essays' in the *St Helena Register*, consolidating and extending the object-lessons in the field for the benefit of the population.[68] No doubt the impending abolition of slavery also focused minds on the true costs of the old practices. Beatson was not the first governor to arrive full of missionary zeal, but many of their initiatives languished as the island imposed its own inertia on their efforts.

One of the more imaginative schemes for boosting the island's natural economy involved the transplantation of cactus plants (*Opuntia* sp.) from the Company's nopalry at Madras (to which they had earlier been introduced from Central America), with a view to breeding cochineal beetles for the production of dye. A letter of 15 June 1791 from Richard Molesworth to Banks mentions that he had heard from James Anderson, founder of the Madras nopalry, that the plants sent from there to St Helena had arrived in good condition, but that initiative too soon foundered.[69]

---

eighteen.' He estimated that of some 1,200–1,600 acres of land fit for tillage, only 120 acres had yet been ploughed.

[67] It may be noted that by this time a large number of the original planters' plots had been consolidated into the ownership of (or were leased by) a smaller number of proprietors. This entrepreneurial class might have been expected to be more receptive to the lessons of the improvers than the original smallholders operating at little more than subsistence level.

[68] Beatson 1816, ix–x. The essays in question, on every aspect of agriculture and animal husbandry, are reproduced in Beatson's volume of *Tracts* (1816). Burchell records the arrival of a further seven Sussex farmers (whisked straight from the East Indiaman *Ocean* to the governor's house on their arrival) in 1808, testifying to the persistence of Beatson's campaign (Castell 2011, 165).

[69] Anderson's role in furnishing the island with new species has perhaps not been fully recognized. Francis Duncan (1825, 162) describes him as a 'constant benefactor' over a period of forty years: 'By the care and assiduity of this excellent person, many rooted plants and seeds of the most useful trees and shrubs, which grow in the EAST INDIES, were forwarded by the Indiamen to ST HELENA; and had the means of the Society enabled them to keep pace with the zeal and ardour of their coadjutor, this island, besides deriving an increase of accommodation from the growth of its plantations, would shortly have become the conservatory of many of the most valuable productions in the world.'

Population and environment: asserting control

Under Governor Walker a plan was drawn up in association with Dr Archibald Arnott, the island's medical examiner who was appointed also as superintendent of gardens, for the introduction of a silk industry. When the first batch of silkworms arrived on the East Indiaman *Repulse* on 17 March 1825 all but three of them had perished on the voyage, but the Company remained optimistic. The estate of The Briars was purchased (see Plate 21) and its notably attractive gardens were cleared away to make way for thousands of mulberry trees; staff were appointed and buildings and machinery installed at a total cost of £20,000. McCracken describes the outcome of the scheme in the following terms:

> It was more than unsuccessful. It was disastrous. Between 1826 and 1834, the venture netted 336¾ lbs (153 kg) which sold in London for £300. It is therefore little wonder that the 1834 commission recommended 'the total abolition of the silk establishment, and the sale of the land and houses and (if practicable) of the machinery, utensils, cocoons, and stock in hand'.[70]

It might be mentioned that the initiative seems to have been aimed specifically to give employment to the newly emancipated black population, so that its failure had a major cost in human as well as financial terms.

Fruit trees continued to blossom in sheltered positions – notably in well-watered valleys – with guavas and figs added to those introduced as early as the 1500s. An oblique glimpse of a further gardening initiative is provided by a record from 1676 of a woman undergoing a public ducking for 'prostitut[ing] her selfe ... A mongst the physick nutt trees'.[71] Cherries seem to have met with little success, while 'gooseberry and currant bushes turn to evergreens, and do not bear fruit'; gardening on the whole seems to have been attended with more uncertainty than in the northern hemisphere, 'the expectations of the gardener [being] frequently disappointed ... his labour is defeated, and his crops often fail'.[72] Governor Pyke planted coconuts which had been sent from Bombay, and several attempts were made to grow coffee.[73] In 1792 Captain Bligh brought with him not only breadfruit (Chapter 6) but also sago and mountain rice. At some unknown point olives were introduced and were said to flourish in the area around The Briars: Dr D. Morris from Kew observed in 1893 that 'extensive areas at similar elevations might be placed under cultivation in olives, which are admirably adapted to the circumstances of the island', but no such initiative was ever acted upon.

---

[70] McCracken 2022, 190–91.
[71] Stern 2007, 13.
[72] Brooke 1808, 18.
[73] The first sack of coffee beans successfully grown on the island was sent to the EIC directors in 1792 (McCracken 2022, 78). When Napoleon was served the local brew he suspected that an attempt had been made to poison him, but later it evidently won a prize for its quality at the Great Exhibition of 1851 (Mellis 1875, 280), though no industry developed for its production.

An interlude of special interest – though ultimately doomed to failure – opened with the arrival in 1689 on board the East Indiaman *Benjamin* of Stephen Poirier and nine other 'vineroons', all of them Huguenot refugees from France accompanied by their families, recruited by the directors of the Company with a view to 'the planting of Vines, and the making of wine and Brandy, which all men of what quality so ever, that ever were upon the Island ... do unanimously agree to be a feazable attempt'.[74] The arrangement seemed propitious, both for the immigrants in fulfilling the Edenic aspirations that inspired so many in the Huguenot diaspora[75] and for the directors who perceived that they had secured the services of a former landed gentleman, whose estates had produced 'two or three hundred hogsheads of Wine and Brandy a year' before being forced into exile. In 1690 members of the pioneering party under Poirier, now designated 'Supervisor of all the Companyes Plantations, Vineyards and Cattle', were duly allotted land on which to establish themselves; when William Dampier passed through a year later he recorded that on the island, 'They do already begin to plant Vines, there being a few french Men there to manage that Affair ... in hopes to produce Wine and Brandy in a short time.' Unfortunately, their optimism proved ill-founded, although the reasons are obscure and were probably complex: lack of resolve, unsuitable soil conditions and the island's rapacious rats have all been singled out for blame, to which climatic turbulence has recently been added.[76] The majority of the islanders meanwhile seemed content with the crude arak they distilled from their more dependable potato crops. By 1695 some of the Frenchmen had lost heart and returned to Europe, but Poirier toiled on: in 1698 he was admonished for his continuing lack of success by the directors, who resolved at the same time to send him further vines which they thought might do better. By the early years of the following century a few barrels of wine were duly sent to London but that was all the produce that ever was gleaned from the initiative.[77] Poirier meanwhile had been given a seat on the island's council at his appointment and went on to serve as governor of the island during a period of particular unrest, when his French ancestry made him a target for local xenophobia – a factor that may further have deflected his attention from attending his vines.[78]

[74] See Court of Committees, 10 August 1688, IOR B/39, f.139: 'ordered that 20 French-men well versed & skilled in planting of Vinyards be sent for St Helena & employed in setting of Vines & that Mr Poirier be made their Overseer: and that such materials be provided for that works as shall be found necessary'.
[75] Stanwood 2020, 6.
[76] A year of drought in 1698 and storms in 1701, 1705 and 1706, as noted by McCracken (2022, 23), must also have contributed to the frustration of many of Poirier's ambitions.
[77] EIC to St Helena, 11 January 1709, IOR G/32/1.
[78] In return Poirier posited a direct link between the degradation wrought on the island through deforestation and overgrazing, and the lax morality of the population – prof-ligacy that he compared with the biblical example of Sodom and Gomorrah – and saw divine retribution in the series of droughts that had hit the island in recent years (see Winterbottom 2010, 208). Clearly the island paradise of a century earlier was already recognized as having been definitively lost.

Population and environment: asserting control

Another purposeful initiative of the Company involved the introduction of sugar cane, which had proved highly profitable under the slave economy of the West Indies. A Mr Cox, formerly resident there, was sent to St Helena with a view to supervising the planned plantations and refineries, but eventually he was dismissed without having achieved success. The sugar cane, at least, established itself on the island: by the 1820s Webster found it thriving moderately there – under the care not of the Company but of the Horticultural Society.[79]

Despite the introduction of all these 'improvements' in plants and animals, the island's capacity to produce the volume of rations expected of it by the Court of Directors was never achieved.[80] Feeding the resident population was one thing, but when a convoy of shipping assembled there awaiting escort to England, the population more than doubled and there was never enough flexibility in the system to feed everyone from the land. Consequently, recourse was made to the salt pork and beef stockpiled in the Company's storehouse – always a popular choice even with the natives, who were able to buy foodstuffs from the Company at below the market rate – an arrangement that must have had the effect of continually blunting their eagerness to develop an enterprising economy of their own.

The introduction of direct rule from Whitehall in 1834 saw the end of the era of such moderate prosperity as the EIC had brought to the island. The new governor, Major-General George Middlemore, appointed in that year arrived with a harsh, cost-cutting brief that impoverished the entire population, left many former servants of the Company in a state of penury and stifled all initiative. All appeals for support from the government were turned away; a decade of petitions to the EIC eventually produced some miserable pensions for its former officers and employees, but the economic devastation wrought during this period by political expediency continues to rankle on the island, while the previous two centuries of tentative progress are regarded, despite their many setbacks, as something of a golden age. The settled population now ebbed away markedly as the prospect of a better life at the Cape lured whole families to emigrate.

Some initiatives continued to be mounted from amongst the (increasingly professional) naturalist community in London, for whom the island held a particular interest. In the 1860s, following the advice of J. D. Hooker, an attempt was made to introduce the cultivation of cinchona in mountainous areas, for the production of quinine. A gardener was sent out from Kew and soon 10,000 cinchona plants were growing on the side of Diana's Peak and promised great success. A pamphlet published by J. H. Chalmers (described as 'Superintendent of Cinchona Plantations and Public Nursery') implies that they met with some initial success. Two species

---

[79] Webster 1834, I, 373.
[80] Commenting on 'the poor crops obtained, and the general barrenness of the fruit trees' even in the later 1800s, Mellis (1875, 42) observed that 'it will not be remedied so long as the existing system, by which all of the vitality of the Island is drained away, remains unchecked. The whole of the manure, which accumulates from stables, stockyards, &c., in the town, is thrown into the sea, instead of being conveyed up the hills, and returned to the land.'

were experimented with under Governor Elliot – *C. succirubra* and *C. officinalis* – but under his successor, Admiral Patey, who was unable to see any advantage of the undertaking, the plantation was neglected and ultimately abandoned.

In the late nineteenth century a flood of new species was further introduced through the agency of the Royal Botanic Gardens at Kew,[81] but none of them flourished as successfully as New Zealand flax (*Phormium tenax*), first planted in 1874. For a few years thereafter the Colonial and Foreign Fibre Company's mill, established at Jamestown, processed the harvests, before closing as uneconomic in 1881. A second attempt at forming a viable industry in 1907, directed by an experienced supervisor (himself imported from New Zealand), proved more successful, and fortuitously the outbreak of World War I boosted demand considerably. Flax quickly came to form the basis of the island's staple industry and a new mill was built at Longwood, after it was found that the difficulties of carting the bulky raw material from the interior of the island to Jamestown rendered that practice uneconomic. Several other processing plants were later built closer to the cultivation sites, which at their maximum covered over 3,000 acres. Some hundreds of growers and mill-hands gained a living from the industry and indeed Gosse observes that so entirely did they give themselves up to it that they neglected the cottage gardens that had sustained them in favour of buying imported food. Such was the appetite for flax-growing that even on the steep sides of the central ridge, 'the ruthless and rapacious flax growers … hacked down and grubbed up wild olive, tree ferns, cabbage trees, lobelia, and everything else which God planted there'.[82] So buoyant was the trade in flax that in 1951 the value of the island's exports exceeded its imports for the first and only time. Then in the 1960s the rise of man-made fibres drove the yarn and tow it produced from the market: the industry collapsed and the populace found itself impoverished anew. The island's economy has been in decline ever since the loss of the flax trade, but today *Phormium tenax* flourishes as one of the most aggressively intrusive species among the wild vegetation, unregarded commercially and subject to culling only in the name of environmental conservation (Plate 11).

For a time the prospect was raised that the island might become a recognized base for vessels engaged in the South Whale Fishery, which called from time to time on their way to and from the Antarctic. At certain seasons the waters around St Helena itself could be enormously productive, leading to the prospect of more local exploitation: one observer in 1805 wrote that

> Whales are seen playing about the island in such numbers that it is supposed the Southern whale fishery might be carried on here with great advantage, as it certainly might with safety and without difficulty, in seas which are never obstructed with ice, nor ruffled with hurricanes.[83]

---

[81] Cronk 2000, 19.
[82] Gosse 1990, 379; Cronk (2000, 19) describes the 'colossal impact' of this movement as 'opening the last act in St Helena's environmental tragedy'.
[83] Quoted in Hearl 2013, 92.

Fig. 10   Boer prisoners of war employed in the whale oil industry, photographed with their try-pots in which blubber was melted down, at Rupert's Bay (1900–02).

Whales were said to be 'numerous in July, August, and September … very large but seldom taken'. Governor Brooke evidently formed particularly good relations with the backers of this trade and discussions were entered into on the possibility of St Helena becoming a depot at which the whalers could leave their catch for transmission onwards to England while they returned to the fishing. The Company decided it wanted no part in the trade, but was willing to sanction its being supported by private enterprise (although at the time there were no buildings capable of housing it). In 1833, £1,000 was raised by subscription towards the scheme,[84] but it came to nothing, as did an attempt to launch a St Helena Whale Fishery Company (Plate 12), which also quickly collapsed. In the later years of the nineteenth century, evidence emerges of much activity by New England whaling crews in the seas around St Helena,[85] from which the islanders benefited economically by servicing ships and crews, though not directly from the industry itself. Attempts to establish an island-based industry continued sporadically (Figure 10) until the outbreak of World War I, beyond which no more is heard of it.

[84] Gosse 1990, 300.
[85] In 1855 forty-three American whalers were recorded at the island. In 1871 some 50,000 barrels of sperm whale oil 'worth at the lowest average £300,000' were estimated to be extracted annually by the American whaling fleet, but to little benefit to the island (Hearl 2013, 96).

Given these immense riches in which the surrounding seas abound, it seems extraordinary that at the end of the nineteenth century the capacities of the island's fishermen to exploit them seem not to have developed one iota since the 1600s. A major limiting factor was certainly lack of investment: there never was a shipbuilding industry on the island and no one had the resources to invest elsewhere in seagoing boats that would have allowed the fishing banks all around the island to be exploited at scale. As a result, inshore fishing with rod and line remained the only method of supply. Governor Beatson's observation that 'proper fishing boats like those used at Brighton' could anchor at the banks and send their catch in daily by smaller craft went unheeded; in 1883 Dr Morris reiterated that the boats available were too small, 'and on this account chiefly many valuable fish found on the windward side of the island are seldom caught'. It seems clear that there was also amongst the fishing population a deep-seated lack of initiative, for a decade later Governor Grey-Wilson reported that it was common practice that 'when a boat has caught a pre-determined quantity, she leaves the fishing ground'.[86]

A report in 1895 from St Helena Industries Ltd not only foresaw profitable trading opportunities for fishermen, particularly with the USA and South America as well as West Africa, but produced full costings for everything from four boats of 30 tons each with their crews, down to iron hoops and rivets for casks. That too came to nothing, but from the early years of the twentieth century the philanthropic Alfred Mosley attempted single-handedly to establish a cannery at Jamestown, with the intention of providing islanders with the means of harvesting the fish (especially mackerel) and selling them – this time to Europe in particular. New boats, fully equipped with fishing gear, were finally brought to the island and stocks of empty cans were imported, but no sooner was everything in place than the stocks of fish that normally abound in the surrounding water disappeared. They continued to be elsewhere for the following nine years, at which point this scheme too had to be abandoned.

Controlling the natural economy

An awareness of the precarious position of the island's ecology had been manifested at an early stage, notably under the enlightened governorship of Captain Roberts. The governor and his council had undertaken a series of surveys in the field, one of which had brought them on 16 October 1716 to the Great Wood. They found it 'in a flourishing condition, full of young trees, where the hogs (of which there is a great abundance) do not come to root them up'. However, they also observed that:

> the Great Wood is miserably lessened and destroyed within our memories; and is not near the circuit and length it was. We believe it does not contain less now than fifteen hundred acres of fine wood land and good ground there are no springs

---

[86] St Helena Industries 1895, 3–4.

Population and environment: asserting control

of water but what are salt or brackish;[87] which we take to be the reason that this part was not inhabited when the people first chose out their settlements and made plantations: but ... if it were enclosed, it might be greatly improved; but doing that would require many hands, the stone, most of it being to be brought a good distance. – but the ground being near to a level for above five hundred acres of it, carts may be used. The enclosing the whole, we think, would be too great an attempt to begin at once; yet we think nothing more proper than to enclose some of the best part, for when once this wood is gone the island will soon be ruined.

An initial solution to the protection of the woods and the crops grown on individual smallholdings was sought in a campaign of enclosure, either by the stone walls envisioned by Roberts or by timber fencing. Finding themselves disadvantaged by being denied access to the forest timber, a number of the principal inhabitants suggested that in return for the right to continue to cut timber, 'Every Planter Possessed of Twenty Acres of Land, Shall be obliged to Enclose one Acre and plant it with wood, and so Proportionably for more or Less': Roberts agreed, on condition that they enclose and plant one acre in ten rather than twenty; a law was duly passed to this effect, specifying not only the nature of the fencing to be used but a maximum distance of 7 feet between adjacent trees, while providing for restocking after felling with 'wood of the same nature'.[88]

The programme of enclosure was continued under Roberts's successors, part of it given over to developing pasture, sufficient to maintain for nine months of the year the Company's stock of black cattle. The remainder of the area (rather less than half the total of about 150 acres) was reserved for the woods that showed signs of continuing regeneration; outside the enclosed area, however, it had again become apparent by the 1770s that no regeneration would be possible while livestock roamed the area in an uncontrolled manner.

An earlier attempt at extermination of the feral sheep and goat population, agreed at a general meeting of the inhabitants in 1731 and enforced by law for a period of ten years (leaving limited numbers to each landowner on condition they were confined), proved remarkably successful in the short term: trees had sprung up spontaneously and woods were re-established in some parts, but evidently

---

[87] Despite the large volumes of water issuing from St Helena, a considerable quantity of it proves brackish, adding to the need to carry the more wholesome streams over long distances for irrigation or human consumption. A further problem was encountered by Halley in 1700, when he called in the *Paramore* to replenish his water supplies: 'the continued rains made the water soe thick with a brackish mudd, that when settled it was scarcely fitt to be drunk' (Thrower 1981, 307).

[88] St Helena Records, quoted by Ashmole and Ashmole 2000, 138–39. The same authors illustrate (plate 8) a stretch of Roberts's wall surviving at Bilberry Field Gut: the bare, eroded landscape it traverses bears witness to the ultimate failure of his imaginative project. Roberts's planting of 5,029 gumwood trees and 500 lemon trees in Plantation Valley in a single year (1709) clearly shows that no blame can be attached to him for the long-term failure of others to capitalize on his schemes.

enforcement gradually lapsed and the gains made were soon lost – not least due to the failure of many owners to confine their stock, preferring to allow them to range freely and fend for themselves.[89] Annual 'pounding days' were instituted when flocks were brought together, animals were singled out for consumption and kids were identified with their owners' marks before being released back into the marginal 'goat ranges'.

Even when the ruination caused by the goats was recognized on the island and measures suggested for their control, the Court of Directors in London proved blind to the problem. Beatson quotes a report from the governor and council of 9 July 1745:

> Finding that great quantities of ebony trees, which grew in or about Peak Gut, in their tender growth, were barked and destroyed by the goats that ranged there, we thought it in your Honors' interest, for the preservation of the wood, and the welfare of the island, to order the goats there to be killed.[90]

The Court replied to the effect that 'The goats are not to be destroyed, being more useful than ebony', ensuring that the problem persisted. Some local initiatives evidently were taken to limit their numbers, especially in view of other hazards they presented. Describing his ascent of the road on Ladder Hill in May 1775, George Forster recorded that,

> Many rocks ... hang over the road, and sometimes roll down to the terror and great risk of the inhabitants, being frequently detached by goats, which come to brouze there; but the soldiers of the garrison have received orders to shoot those animals as often as they appear on these eminences; and no other command is obeyed with greater alacrity, because they are generally permitted to feast upon the goat which they have killed.[91]

The misplaced influence of the distant Court of Directors condemned the island to generations of continuing depredation of the botanical stock and to the disastrous soil erosion that followed as its inevitable consequence. When he came to publish his *Tracts*, Beatson was typically forceful in expressing his view: 'I am perfectly aware of the arguments adduced in favour of the goats – I have weighed them maturely, and I am thoroughly convinced the whole are nugatory'. The

---

[89] Evidently the planters themselves represented almost as big a menace: only three years later a survey by the governor found that over 1,000 young trees in one of the Company's plantations had been deliberately cut down and carried off, a pattern of destruction repeated all over the island (Gosse 1990, 168).

[90] This measure had been mooted earlier under Governor Roberts and (more successfully) under Governor Byfield, who succeeded in convincing some of the landholders that the move would be in their own interest; although Byfield enjoyed some success in regenerating the forest in this way, his achievements had evidently been dissipated in the following decades.

[91] Forster 2000, II, 662.

Population and environment: asserting control

extermination of the goats and a reduction in the number of sheep he considered an essential step, without which 'there can be no hope of ever rendering [St Helena] a valuable property to the Company'.[92] Remarkably, even after such a trenchant (and wholly justified) assessment of their negative impact, it would not be until the early 1960s that the last feral goats were effectively eliminated.[93]

With the loss of the woods, soil erosion increased at an alarming rate. At the turn of the nineteenth century, Brooke observed that

> the rain has made great ravages, as the soil was deprived of its adhesive quality by the want of that humidity which accompanies foliage and shade. Serious apprehensions were entertained that the evil might become general; and the island of Bermuda, and other parts of the world, were cited as instances to prove that countries highly fertile, when abounding with wood, were reduced to barrenness when deprived of such clothing.[94]

Not all of the damage to the forests can be attributed to the depredations of the feral goats and swine, for with the growth of the settled population a number of trades contributed to the drain on timber resources: in the eighteenth century, ebony was being used to burn lime,[95] while the bark of both ebony and the native redwood was utilized by tanners. The latter were singled out for criticism in 1709 when the loss of both species was said to have been hastened by the tanners, 'that for laziness never took the paines to bark the whole trees but only the bodys, leaving the best of the bark on the branches, by which means has [sic] destroyed all those trees, at least three for one'.[96] But neither should all the losses be attributed to human agency: a letter of 23 February 1786 from Governor Daniel Corneille to

---

[92] Beatson readily acknowledged the value to the economy of properly confined pigs: indeed, he writes, there is 'no species of husbandry so well adapted to St Helena as that of hoggeries', particularly in light of the ease with which yams and other root crops grow on the island. The well-watered valleys would blossom with trees and gardens, he predicted, while 'the steep declivities on either side, would sufficiently protect the plantations from the trespass of black cattle' (Beatson 1816, 6–9).

[93] [Brooke] 1810; Ashmole and Ashmole 2000, 160.

[94] Brooke 1808, 212.

[95] Much of the ebony involved in this process, however, would appear to have come from dead trees: a consultation of 12 July 1709 reported that 'there are huge quantities of that wood lying dead on the hills near Sandy Bay ... and just by that place where the wood lies, are mountains of extraordinary lime stone; and it will be much cheaper to our honourable masters to bring lime from thence, ready burnt, (being light) than to fetch that sort of wood (which is very heavy), and bring it to the castle in James's Valley' (quoted in Beatson 1816, 2). Brooke (1808, 141), however, records that the burning of limestone was suspended within a few weeks, presumably when the scale of devastation it occasioned became visible.

[96] Mellis 1875, 226–27. Brooke (1808, 139–41) mentions that ebony bark was exported to tanners in England and the West Indies, so it was not only the local industry (which at first was encouraged then restricted by the Company in 1709) that benefited from these depredations on St Helena.

Joseph Banks explains his failure to send a specimen of ebony, citing the fact that 'all the trees have been destroyed by an insect'.[97] Redwoods had reached the brink of extinction when Governor Byfield intervened to stem the loss:

> Five acres of Plantation-House grounds were enclosed within a wall fence, and adopted as a nursery; and the red-wood, which had become nearly exterminated, was by this means preserved. Governor Byfield met with two young plants of it, which were moved into a proper situation, and protected till they produced sufficient seed to multiply their numbers.[98]

Brooke makes an interesting statement that under the mismanaged regime of Governor Roberts's successor, Captain Boucher, not only was the produce of the Company's farms 'wasted by the most wanton extravagance' but that 'a fine herd of deer [was] totally extirpated'.[99] The Plantation House gardens, meanwhile,

> were laid waste, and thrown into pasture for the Governor's asses, of which he kept a numerous stud; and that he might have his favourite exercise of riding them in all weathers, a shed, of four hundred feet in length, was erected, at the Company's expense.

Boucher was probably the least well-equipped governor to cope with a drought followed by an epidemic that wiped out 2,500 head of cattle during his period of office. As mentioned earlier, shortly after Governor Pyke replaced him in 1714, a census showed that there remained only sixty head of cattle, twenty-four swine, three sheep and twenty-six fowls on the island. Pyke ordered that until stock increased, white inhabitants were to eat salt meat two days a week and fish on two others.[100]

At times, flooding could prove equally devastating. Under Pyke's governorship, a year of drought that caused extensive deaths in 1718 was followed the next year by an equally destructive deluge:

> ... supposed to have been produced by the breaking of a water-spout over Sandy Bay Ridge ... It washed away the mould from the mountains, filled up some of the plantations with stones and rubbish, and swept others entirely away. The sea for many miles around was discoloured with mud, many families were nearly ruined, and it required much industry to repair the mischief; but it was the operation of time alone that could restore the hills to their lost covering.[101]

---

[97] Banks 1958, 231.
[98] Brooke 1808, 213.
[99] Ibid., 190.
[100] Gosse 1990, 135. As an interesting sidelight, in 1829 mynah birds were introduced from India in the hope that they might destroy the ticks and other pests that plagued the cattle. The experiment met with only moderate success, but at a second attempt to introduce them in 1885 they multiplied to such a degree as to become pests themselves, causing especial havoc among fruit-farms and gardens.
[101] Brooke 1808, 195. Gosse (1990, 134) quotes another source describing the torrent as 'carrying away the soil in an incredible manner, with both grass, trees, yams and stone

Flooding, droughts and the constant trade winds proved inimical to the introduction of some species. Scotch firs, pines and spruce trees were imported during the middle decades of the 1700s and oaks were successfully grown from acorns. Later the Company sent living specimens of Lombardy poplars, but they failed. At the beginning of the nineteenth century, under Governor Beatson, 11,000 pineaster (maritime pine) saplings raised in the Plantation House gardens were transplanted to sites around the island, and a further 27,000 were planted under Beatson's successor, Colonel Wilks, in 1816.[102] A system of premiums was introduced to encourage landowners to follow suit. The local gumwood flourished more successfully than any of these imports, but throughout this period pressures from the human population proved most destructive. The doubling of the population during the years of Napoleon's incarceration on the island no doubt increased demand far beyond any sustainable level. During this period Governor Hudson Lowe, who assiduously exploited the Plantation House gardens as a nursery for seedling trees, applied both carrot and stick in motivating the wider population, introducing a series of premiums for those who undertook planting and threatening those who cut down living trees in the island's interior for use as firewood with fines of up to £10 for each offence or corporal punishment not exceeding 200 lashes.[103]

As its name might suggest, the estates of Plantation House – the governor's country residence – played host to a number of initiatives aimed at mitigating the threats posed to the island's plant cover. The 500 lemon trees planted there in 1709 formed only one of a series of initiatives set on course by successive governors, all ultimately with indifferent results – like the pineapples introduced there a century later or the redwoods whose careful preservation is mentioned above. The thriving landscape with regular beds recorded during Beatson's governorship by George Hutchins Bellasis (see Plate 10) must represent a rather fleeting high point, for even here there were more disappointments than triumphs and the initiatives of one governor seem often not to have found favour with his successor. There was a grain of truth in the review of the island's history of failed initiatives compiled in the early twentieth century by Colonel Robert Peel, who commented that 'Experiments in Agriculture are to be avoided in St Helena – the people know what will grow and what will not.'[104]

---

walls before it. It brought down rocks of a mighty bulk, and covered abundance of fruitful land with stones.'

[102] Ashmole and Ashmole 2000, 155. Mellis (1875, 222) attributes much of the nineteenth-century decline of endemic species to the energy with which Beatson introduced exotics from all parts of the world: 'these have propagated themselves with such rapidity, and grown with such vigour, that the native plants cannot compete with them … and wherever established, they have actually extinguished the indigenous Flora'.

[103] See McCracken 2022, 123.

[104] Field 1998, 54–55.

All too often, however, the truth was more complicated. A campaign of planting gorse as a measure to reduce pressure on the woods and to provide shelter for livestock (as well as forming an alternative source of fuel), for example, proved an initial success. Instituted under Governor Pyke, it was maintained under Byfield, who ordered annual surveys to ensure that farmers maintained their allocation of gorse planting (along with fence-building and the stipulation mentioned earlier that each should rail off one-tenth of their property in order to plant trees to help regenerate the forest), and enforced quotas by means of fines. Banks said of the gorse in 1771 that it 'Thrives wonderfully and is highly prais'd by the Islanders as a great improvement, tho they make no use of it except heating their ovens'.[105] By the time George Forster visited, however, only a few years later, in 1775, it seems that the initiative had run its course and was now being thrown into reverse:

> The common furze or gorse (*Ulex europœus*) which our farmers take great pains to eradicate, has been planted here, and now over-runs all the pastures ... The aspect of the country was not always so delightful as it is at present; the ground was parched by the intense heat, and all kinds of herbage and grass were shrivelled up. The introduction of furze bushes, which throve as it were in despite of the sun, preserved a degree of moisture in the ground; under their shade the grass began to grow, and gradually covered the whole country with a rich and beautiful sod. At present the furze is no longer wanted, and the people assiduously root it out, and make use of it for fuel, which is indeed scarce upon the island, though I never saw a more œconomical use made of it than here, and at the Cape.[106]

Another initiative that went awry was the introduction of the English blackberry, again with the idea of providing hedges and windbreaks as well as fruit. So successfully did it adapt to its new habitat that by the early 1800s it menaced much of the pasture and arable land. Such was the scale of the problem that the meagre labour force available was deemed incapable of tackling it: Governor Patton's answer was to form working parties from the men in the garrison, to be paid by the individual landowners to clear their land, following which, responsibility for keeping the blackberries in check would revert to them.

The best pasture in the uplands was widely sown with English vernal grass; lucerne was also successfully grown. The native wire grass flourished in the low-lying areas: Brooke describes it as sweet and nutritious, and with a capacity to survive hot and dry weather, but many acres of it were invaded by a coarse and less useful herb, called (ironically) cow-grass. Brooke was able to extend the area of pasture by introducing irrigation schemes, notably by channelling springs on the slopes of Diana's Peak to water the plain at Deadwood.

---

[105] Banks 1962, II, 267.
[106] Forster 2000, II, 665. One wonders whether Forster may have misinterpreted the deliberate harvesting of gorse for attempts to eradicate it.

Population and environment: asserting control

One inadvertent introduction caused perhaps more destruction than any other. In 1840 the timbers from a slave ship intercepted on the southern passage to Brazil were brought into Jamestown and, no doubt, recycled in the usual manner. In the decades that followed 'termites' (more likely white ants) which had been introduced with the timbers began to wreak havoc on every organic substance they could access; houses and public buildings were ruined throughout the town.[107] Its appearance in the late 1860s was described by Mellis in 1875:

> It was a melancholy sight five years ago to see the town, which had hitherto not been without its claim for admiration, devastated as by an earthquake, or, as a visitor remarked, a state of siege – the chief church in ruins, public buildings in a deplorable state of dilapidation, private houses tottering and falling, with great timber props, butting out into the streets and roadways, meeting the eye at every turn … while the Governor in his council-chamber, the Chief Justice, and the other officials, were accessible only through a labyrinth of fir-poles and old ship-planking set on end to prevent ceilings falling on their heads, or, worse still, whole buildings collapsing around them.[108]

Under Admiral Sir Charles Elliot, appointed governor in 1863, a campaign of extermination of these pests and reconstruction of the settlement in stone, iron and teak, resulted in almost all the public buildings being renewed.

By this time the settlement at Jamestown was already severely overcrowded, but prevented from expanding by the steepness of the valley. When Governor Gore Brown was appointed in 1851, despite having been handed yet another mission to cut costs on the island, a decision was taken to establish a settlement at Rupert's Valley. The first major building there, a prefabricated wooden prison, 'was burnt to the ground in less than an hour by a prisoner confined in it in 1867'.[109] The village project was forwarded under the succeeding governor, Sir Edward Drummond Hay, but despite much expenditure in leading water to the site, it proved unviable; today scarcely a trace of it survives.[110] By this time the island was so impoverished that some 1,500 of its inhabitants emigrated to the Cape.

Under Governor Elliot, Drummond Hay's successor, renewed efforts were made to introduce new and valuable species to the island – Mexican pine, Norfolk Island pine, Bermuda cedar – but they all failed due, it seems, to inadequate husbanding. A planting campaign was carried on into the early decades of the twentieth century in an attempt to stem the continuing denudation of the forest cover by islanders in search of fuel.

In the course of the century following the end of East India Company rule, increasing impoverishment among the population led to a gradual consolidation

---

[107] Gosse 1990, 312.
[108] Mellis 1875, 171–72.
[109] Gosse 1990, 321.
[110] Ibid., 323.

of country properties so that in a census of 1931 it was found that 70 per cent of the occupied land was now in the hands of only three owners – the only residents capable of raising the necessary purchase money when land came on the market.[111] Initiatives were drawn up by the Colonial Office to build cottages and to install smallholders, but were never implemented for lack of resources. Many existing country properties were characterized at this time as 'wretched hovels', while even Jamestown harboured many slum properties. Imported foodstuffs and even agricultural fertilizers sold at such high rates as to put them beyond the reach of all but the few prosperous proprietors.

The struggle to find a basis for economic independence continued, but with an undeniable air of desperation. In the wake of the great International Exhibition movement that swept Europe and the rest of the developed world at this time with the aim of boosting trade and industry, St Helena mounted its own Industrial Exhibition in the Castle at Jamestown in 1874.[112] Amongst the 'exportable produce', phormium flax, aloe fibre, coffee, cotton and cured fish were all identified as 'giving promise that a trade in these exports may hereafter be developed', while other exhibits such as cured hams, dried fruits, preserves and pickles, honey and wax, wheat and flour were deemed to 'shew that much of provision now imported may be produced on the Island'. A list of domestic manufactures held out as potentially profitable seems less convincing: cabinet work, straw work, fishing rods, walking sticks, lace (made from the fibre of American aloe), ornaments and fancy articles were nonetheless identified as pointing to 'sources of industry as yet undeveloped'. Even the report of the organizers seems to betray doubt that there was much hope here, beyond seeking 'to encourage the industry of our population and to divert them from that necessity for emigration which has deprived St Helena of many of its best laborers and mechanics'.

A measure of the extremity of the economic situation is marked by the £800 increase in the island's revenue in 1904, generated from sales of the island's Victorian postage stamps[113] (special issues continuing to be valued as a source of income to this day).

---

[111] Mellis (1875, 29) records that the EIC's annual expenditure on the island had been in the region of £80,000 or £90,000 annually, a sum that throws into perspective the size of the loss to the island's economy occasioned by its transfer to an exceptionally parsimonious government.

[112] See *Report of the St Helena Industrial Exhibition*, 1874.

[113] Gosse 1990, 344. The regularity of even this source of income could not be guaranteed: the same author records (p. 351) that a new issue of Ascension Island stamps in 1922 resulted in a £2,000 fall in income for St Helena the following year.

5
'The citadel of the South Atlantic'

The towering impregnability of St Helena impresses itself upon every seaborne visitor to the island, any weaknesses in its natural defences having been buttressed over the centuries by successive campaigns of fortification by the East India Company and later garrisons. William Webster characterized it (somewhat whimsically) in 1829, shortly before it passed from Company control, as follows:

> here art vies with Nature's grandest efforts; fortresses with their turrets and cannon bristle on every point and pinnacle of rock. In fact, the position, the strength and number of the fortified points, appear to denote that the ambition of its possessors would render it the citadel of the world. A more military station I know not, for it far surpasses Gibraltar, and one naturally asks himself, whence is all this solicitude, this unnecessary zeal, and overweening anxiety for its security? – whence the advantage of being safely caged in this island while the sovereignty of the sea confers on it immunity from danger? and when that is lost to us, of what importance or value could be such a rock as this? Such, however, were not the questions of those who planned the mighty works of St Helena.[1]

British 'sovereignty of the sea' must have seemed as secure as the rock itself in Webster's day, but it had not always been so. Before the British established a permanent foothold on the island, visiting seamen evidently had no thought of fortifying it against each other, although they might occasionally have hauled their cannon on shore to provide covering fire for vessels riding at anchor off Chapel Valley. We hear of an altercation in 1625, when a Portuguese (or possibly Spanish) carrack was surprised at its moorings by the arrival of a Dutch ship: some of the cannon were immediately landed and succeeded in beating off the Dutch, whose vessel sank nearby the following day. Having salvaged a good deal of the contents and the fabric, the Portuguese constructed a breastwork on which the captured cannon were mounted and used to good effect when a Dutch fleet of six vessels arrived shortly afterwards and was successfully repulsed by the shore-based bombardment.[2]

The Dutch had proclaimed their aim to fortify and populate the place in 1633,[3] but never actually did so; with the transfer of their attentions to the Cape, it fell instead

---

[1] Webster 1834, I, 340–41.
[2] Gosse 1990, 34–35.
[3] See Chapter 2.

to the British to establish a permanent foothold there. The EIC's title to the island and its capacity to act in a quasi-military manner were given formal recognition (at least in Britain) with the granting of Oliver Cromwell's charter in 1657, and in the following year the directors of the Company resolved 'to send forty men with all expedition to remain on the island with conveniences to fortify and begin plantation there'.[4] Captain John Dutton was named as 'Governour in Cheife on the said island',[5] with a commission from the newly succeeded Richard Cromwell to take possession in the most formal manner (though the ceremony can have been witnessed by no more than a handful of onlookers). The Court of Directors declared that:

> Wee doe therefore require you that ... in the name of His Highnesse Richard, Lord Protector of England, Scotland, and Ireland, and the dominions thereunto belonging, and for the use of the Honourable English East India Company, doe take possession of the island, and with drum and trumpett proclaim the same.[6]

With the Company's third charter, dated 3 April 1661, permission was formally granted by the newly restored King Charles II that the merchants of the EIC might

> fortify St. Helena, and any other places within their limits of trade, and supply their forts and settlements with cloathing, victuals, ammunition, and implements, free from custom or duty, and transport thither such men as shall be willing to go; may govern them in a legal and reasonable manner, and inflict punishments for misdemeanors, or impose fines for breack of orders.[7]

A further grant of 1673 confirmed the Company's rights on the island in perpetuity.[8]

By this time St Helena had become formally a British possession, with a military garrison and the rudiments of a settled population of planters. While interest in the island remained centred on its capacity to act as an Atlantic staging post for East Indiamen on the homeward voyage, its long-term effectiveness in re-victualling as well as re-watering ships was dependent on the establishment of a viable, productive and ultimately self-sustaining civilian society – an ideal that would never quite be realized by the EIC and which remains elusive today.

Securing the island

Soaring cliffs encircling much of the island form natural ramparts of the most spectacular kind: Peter Mundy wrote in 1638 that 'there is hardly such another Ragged, steepy, stony, high, Cragged, rocky, barren, Desolate and Comffortless

---

[4] Sainsbury 1912, 289.
[5] Dutton, who had formerly sailed with the Dutch East India Company, had been designated in 1658 as the first British governor of Pulo Run in the Spice Islands, when the island was ceded to Britain by the Dutch, but had been prevented from taking up the post by a change of heart on the part of the latter.
[6] Foster 1919, 284. Richard Cromwell succeeded his father as Lord Protector on 3 September 1658, ruling until 25 May the following year.
[7] *Collection of Statutes*, appendix, 8.
[8] Ibid., x.

'The citadel of the South Atlantic'

coaste'.⁹ The need for fortification of such a formidable barrier was therefore limited to those few areas where natural valleys broke through the scarp to form potential access points at sea-level.¹⁰ Only one of these, on the leeward side of the island, is of any size: Chapel Valley (renamed along with its settlement as James Valley, in honour of the Duke of York, later King James II) provides the sole dependable landing spot, although even here gaining an anchorage was by no means an easy matter, as recorded by Francis Rogers in 1701.

> as we go to fetch into the road, which lieth to the leaward of the Island, coming about the Northernmost end, [we] luff so near that they hail the ship from a little look-out upon the high rocks … Oft as we pass by the openings of the rocks or open a valley … strong flaws of wind comes off.
>
> The sailors are obliged to stand by their topsails as the ship opens these valleys, to shoot her into the road as near the land as they can, to get the better anchoring ground off James's Valley, where the ships ride in 30, 40 or more fathom water as they are in or out, for 'tis very steep too here, and if a ship drives but a little out of the usual anchoring place, she presently loses the bottom and drives away to leaward, and she may turn long enough before she fetches in again, maybe a week, maybe not at all.¹¹

Otherwise, on the (eastern) windward coast, Prosperous Bay and Sandy Bay form possible landing places, although heavy seas there present a hazard to vessels approaching the shore and provide an additional degree of defence.

Even today, there is no harbour at Jamestown (although a recently constructed jetty at Rupert's Bay has eased access to the island: see Chapter 9). Larger ships still anchor off in the roads and send in tenders, whether to deliver passengers and freight or to take on water. As well as repelling unwelcome attempts to gain the shore, the defences established there had the dual role of providing covering fire for shipping at anchor. The sheltered anchorage compensated adequately for the lack of docking facilities, however, so that considerable numbers of vessels might be found there on a seasonal basis, waiting to join an armed convoy for the final leg of their voyage home.¹²

---

9   Ólafsson 1932, III, part 2, 412.
10   Even in Governor Beatson's day, it was recorded that 'out of twenty-eight miles of coast, the fortified lines of defence, collectively, do not exceed eight hundred and fifty yards' (Beatson 1816, lxxxvii).
11   Rogers 1936, 191. Van Linschoten (1885, 249) had earlier alluded to the difficulty of finding a berth there, warning that approaching mariners must always keep the island to port: 'otherwise it were impossible for them to come at it if they leave that course: for if they once passe it, they cannot come to it again, because there bloweth continually but one kind of wind, which is South east'. Tavernier (1678, part II, 207) gives similar advice.
12   Non-EIC vessels were not welcomed: diplomatic *politesse* might occasionally demand that certain foreign ships be accommodated (although their crews were allowed ashore only in small numbers, unarmed, and closely monitored to deter spying), but

Defences and defenders

The securing of the island had been stipulated as a first priority for the new governor:

> The Allmightie having arrived you at St Hellena, you shall with all speed proceede to fortifie in the most convenient place of Chappell Valley ... and in such other place and places on the island as you shall judge necessarie and requisite for the defence of the same and to offend any enemies that shall come into or neare the roade or roades of the said island.[13]

The earliest and most prominent fortification on the island was the triangular fort with three bastions erected at the foot of the valley under Governor Dutton in 1659 (Figure 11); it was named Fort James, and with the decline of the chapel established on the site by the Portuguese, the fort and its attendant settlement, the embryo Jamestown, now gave their name to the valley. The Sieur de Rennefort, visiting the island in 1666, mentions the presence of twenty cabins in the interior of the fort, serving as barracks for the garrison.[14]

This and the other early fortifications clustered in the north-west of the island (Figure 12) were put to the test when in 1673 the Dutch temporarily wrested control of the island in what can have been little more than an incidental engagement of the Third Anglo-Dutch War of 1672–74.[15] The task evidently was accomplished with comparative ease, and although conflicting accounts survive, English and Dutch, of precisely what took place, both are agreed that the action involved only four men-of-war belonging to the Dutch East India Company (VOC). According to the English record of events, the Dutch appeared off Jamestown around 20 December 1672 but were driven off by fire from the fort and from the Company ship *Humphrey and Elizabeth*, then anchored in the roads; a sustained assault was then mounted at another place (possibly Lemon Valley) before the besiegers mounted a successful landing of 400 men at Old Woman's Valley; the defending force of 170 fighting men found themselves overwhelmed and were forced to retreat, first to the fort and then to the *Humphrey and Elizabeth* and another EIC vessel that had appeared in the meantime; at this point, having spiked the guns and carried away all the military stores they could handle (plus the women and children), they abandoned the island and set sail for Brazil. The Dutch account leaves less scope for the salvaging of any reputation by the defenders: it has the squadron arriving nine days later then the English report and, after a delay caused by contrary winds, landing 300 men virtually unopposed at Old Woman's Valley;

---

      in general even those from England not sailing on behalf of the Company were rebuffed (Royle 2007, 128–30).

[13]   Foster 1919, 285–86.
[14]   Gosse 1990, 52.
[15]   It was, however, a purposeful operation, for the Dutch had sent the ships first to the Cape to embark soldiers and then to St Helena to execute their purpose. By the time they sailed from the Cape, on 13 December, they carried between them 634 soldiers and 110 guns. See Kitching 1950, 58–59.

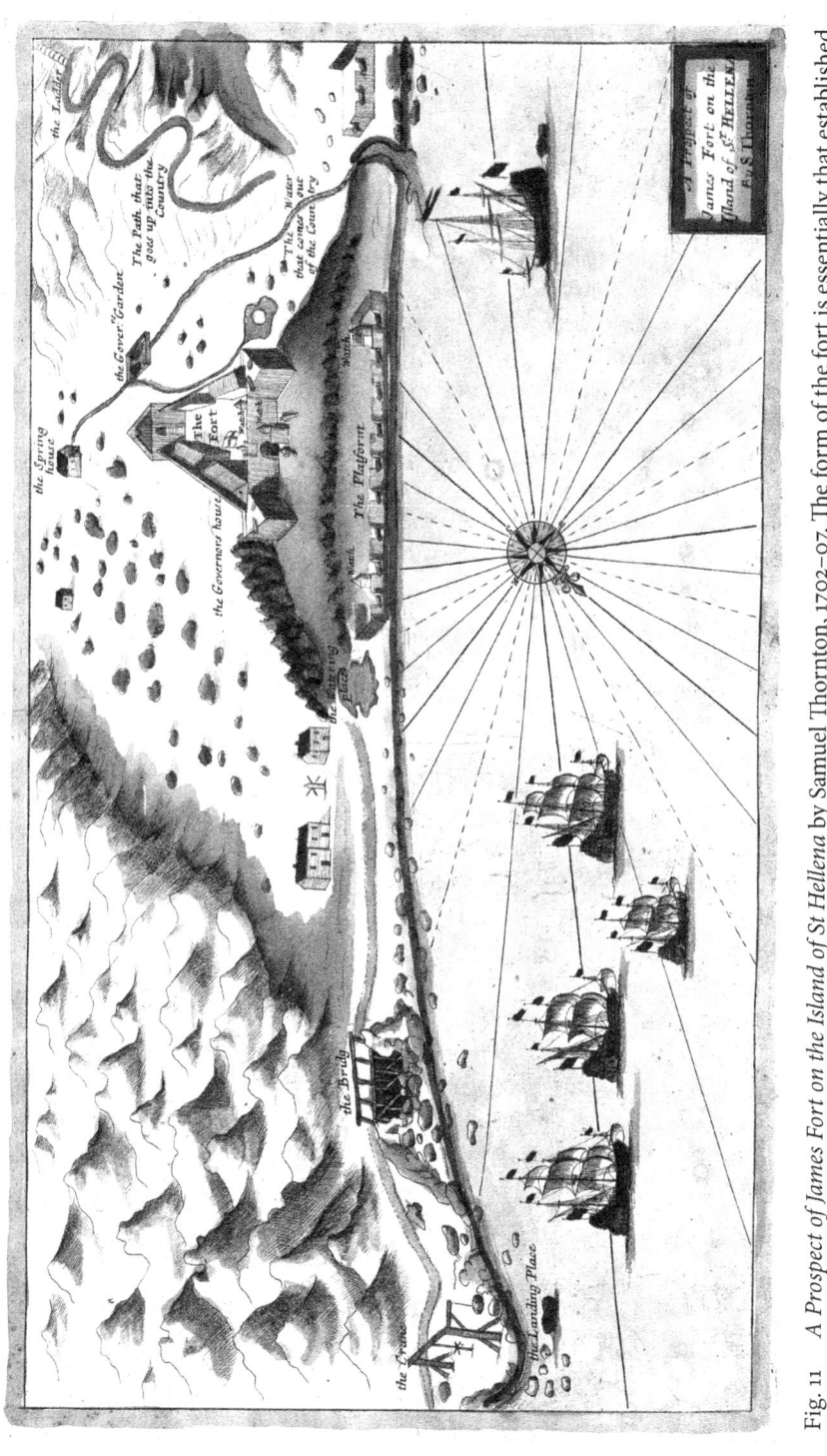

Fig. 11 *A Prospect of James Fort on the Island of St Hellena* by Samuel Thornton, 1702–07. The form of the fort is essentially that established by Governor Dutton in 1659.

with the aid of one willing collaborator they advanced to Jamestown where they found the fort already abandoned. Although it was convenient for the Company to endorse initially the more favourable account, no one could fail to acknowledge that the defence of the island had failed miserably.[16]

Fortunately for the EIC, the defensive shortcomings laid bare by this episode worked to their advantage the following summer when the Royal Navy frigate *Assistance*, under the command of Captain Richard Munden, chanced across a small schooner sailing to the east of St Helena.[17] On board were the displaced governor of the island, Anthony Beale, with some crew members from the *Humphrey and Elizabeth*: Beale had chartered the vessel in Brazil and had taken it upon himself to cruise to windward of the island in order to intercept English vessels that might be heading there, unaware that it had fallen to the Dutch. Hearing of the island's plight and having, in his own words, 'noe other business to doe', Munden resolved to recapture the island. The *Assistance* was carrying a considerable number of men at the time, some of them soldiers sent to reinforce the island in anticipation of hostilities reaching that far; others came from amongst the crew of a dismasted ship that Munden had taken on board, together with their master, Captain Keigwin.

Munden landed 350 men under Keigwin at Prosperous Bay, at a spot now known as Keigwin's Point, before beginning an assault from the sea; anchoring close to the shore at Jamestown, he bombarded the defences there and the Dutch (whose garrison numbered only about 100 men) quickly concluded that they would be unable to maintain their resistance. They offered their surrender to Munden, who sent ashore in return the captain of one of his support ships 'with ye King's colours to take possession of ye fort' and with a trumpeter to signal to the landing party that the island had been secured; they entered the port peaceably the following day. The whole of the 'reconquest' had already taken place by the time news that the island had been lost to the Dutch had reached London. Munden was rewarded with a knighthood and with gratuities of £2,500 by the state and £400 by the Company; lesser sums were paid to the other commanders, including Keigwin, whom Munden left as acting governor of the island.

These details are summarized from Stephen Royle's wonderfully detailed chapter on 'Defence and the imperial imperative' in his volume *The Company's Island*. Here the location of the fortifications and lookout points is mapped and the allocation of men and resources to the more vulnerable coves around the coast

---

[16] Royle 2007, 146–50. The failure of the two Company ships and one French vessel present, mounting 100 guns between them, to engage the Dutch squadron, whose four ships had been left in the roads with only 130 men between them, also came to attract criticism.

[17] The *Assistance* had left England in January 1673 as escort (with three pressed vessels) to a convoy of ten East Indiamen heading first to St Helena; from there the merchantmen were to proceed to the east, while the escort was to accompany another convoy on its return voyage to England.

'The citadel of the South Atlantic'

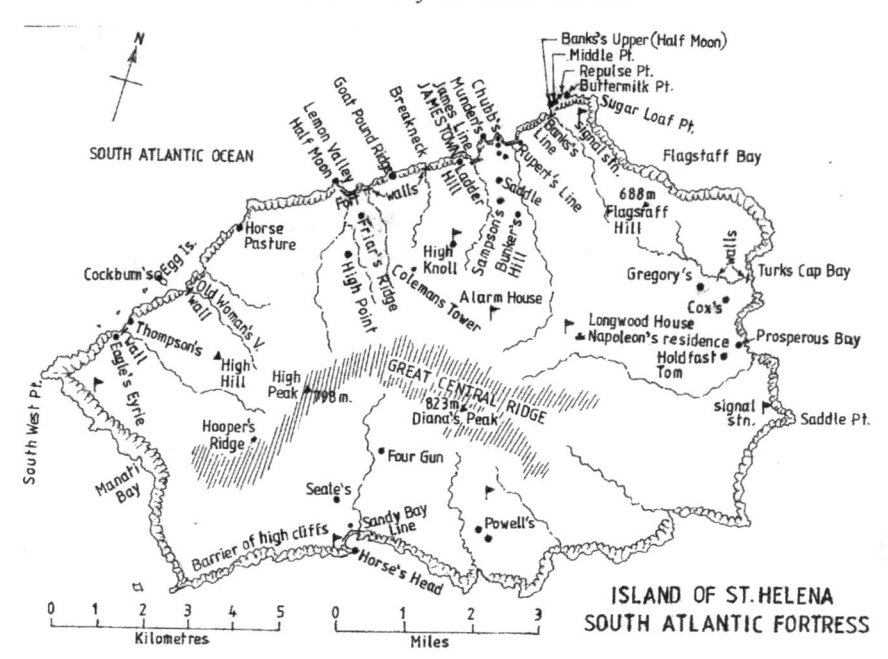

Fig. 12   Map of the principal fortifications and batteries on St Helena. From Ken Denholm, *An Island Fortress* (2006).

and to their associated defences is discussed at length. Armed response at different sites varied from heavy cannon (sometimes requisitioned from passing shipping) to crowbars with which rocks could be dislodged to rain down on intruders.[18] Royle also reproduces the text of a brief survey of the defences compiled by a visiting ship's captain, William Bass, on 25 April 1674:

> Near ye Sugar Loaf a platform built for 5 gunns with a guard house and a place for ammunition but upon it no more but 3 gunns. It goes by ye name of Bank's Platform. In same valley a platforme for 7 guns built but upon it but 5. It goes by ye name of Rupert's Platform. At ye next towards Chapel Valley is a platforme and 2 gunns on it goes by ye name of Munden's Point. In that which was formerly called Chappil now James Foart a platform of 16 guns, at ye crane or landing place 3 guns on ye two bastions, In ye fort 5 guns. Upon parsley bed ridge 2 guns at ye head of ye same valley to alarum ye people when any ships was seen. At Lemon Valley known by ye name of Springg's Plattforme 5 guns with a guardhouse and

---

[18]   Although the measure sounds primitive, a few men could put up a devastating defence by rolling rocks down the narrow ravines and valleys. Seventy-six crowbars (also used to manoeuvre the cannon) were inventoried in 1701. Gunpowder charges might also be used to topple rocks on an invader (Royle 2007, 144–45).

Ammunition house. Three of those guns are on ye sides of ye hills and ye other two on ye platforme below.[19]

By 1691 the experienced eye of William Dampier found the defences around the bay at Jamestown 'so strong, that it is impossible to force it'. Even the cove where Munden had staged his reconquest, though 'scarce fit for a boat to land at', he found had now been fortified.[20]

New regulations were also issued for the organization of the garrison, which was now to be formed into two companies, one under the governor, Captain Field, and the other under his deputy, Captain Beale. Regular exercises were devised to develop expertise among the junior ranks.[21] Soldiers and planters were to collaborate in constructing fortifications (at times assisted by the crews of passing ships and increasingly bolstered by slave labour) and in keeping watch.[22] A condition of the grant of land to planters was that they should serve in the militia; new regulations decreed that they should be 'taught and instructed in the exercise of Armes'; they were to turn out whenever the alarm was raised by signal flags or by drumbeat (and later by an 'alarm gun' positioned on the hillside above Jamestown). Conversely, it was anticipated that soldiers would play a role in food production (and indeed in the earlier years of settlement most were billeted among the civilian population); they were not allowed to own plantations while on active service, but numbers of them became planters on their retirement. In the decade following the Dutch invasion, the garrison fluctuated between about fifty and eighty men.[23]

Whereas the militia raised on the island was comprised primarily of riflemen, the guns in the fortifications were manned by the Company's troops – irregulars forming the first generations of what would later develop as the more formally constituted St Helena Artillery, staffed by officers and men recruited and trained through the Company's regular service. The progress of the St Helena Infantry followed a similar path, although its reputation seems to have been perpetually blighted by discontent and indiscipline in the ranks, periodically boiling over into mutiny. The most serious of these incidents, in 1787, concluded with a full-scale fire-fight, in which 103 mutineers were eventually taken prisoner and 99 of them sentenced to death; in the event it was decided that only one in ten would be shot, the choice to be determined by the drawing of lots.[24]

[19] IOR l/MAR/A LXXI, 22–27 April 1674.
[20] Dampier 1937, 364.
[21] IOR E/E/3/88, ff.285v–89v, 15 March 1678.
[22] Ships arriving off Jamestown undetected by the lookouts would result in fines being imposed for laxity. Later stick was replaced with carrot, when Governor Roberts introduced the practice of rewarding the first person to spot a ship with one dollar (Royle 2007, 146).
[23] Royle 2007, 132–34.
[24] Gosse 1990, 206–10. The only shots fired in anger in the history of the garrison were indeed in the cause of suppressing periodic mutinies amongst both the civilian and

One activity that the Company sought to rein in at an early date was the over-liberal firing of salutes between visiting vessels and the shore batteries and the consequent drain on supplies of gunpowder.[25] A set of orders and instructions from 1 August 1683 complains that, 'Wee finde, by the list of guns fired, sent us by Capt. Beale, three hundred and odd guns, which is so strange a waste, that we could not think our Governor would have bin guilty of'. The directors took particular exception to the saluting of 'interlopers' – non-Company (and often non-British) vessels that the directors held had no business to be calling at the island and which they were reluctant even to see supplied with water.[26]

In 1703 Governor Poirier was exhorted to 'give no rest to your thoughts and endeavours till the island be made not only tenable against the invasion of any enemy whatsoever but also able to defend our ships in harbour there'; his responses proved woefully inadequate, for three years later Poirier had the mortification of losing two Company ships to the French as they lay at anchor in the roads.[27] Impressive as the batteries may have appeared, their effectiveness was dependent on due observation of procedures, both on land and at sea. On this occasion the ships lost to the French had for their part stood off from the shore too far to benefit from covering fire from the land; the batteries were indeed ordered to open fire, but apart from the problem of range it was found that 'a sufficiency neither of powder nor match was at hand and many of the spunges did not fit the guns'. Reforms followed on the inevitable recriminations, resulting in new orders for shipping to moor close to the shore and for none to pass Banks's Battery without sending a boat ashore to establish their credentials.[28]

Poirier's successor as governor, Captain John Roberts, undertook the reforms to the defences that Poirier had signally failed to carry out. Brooke mentions that on the day he arrived, Roberts – perhaps the most effective governor of the eighteenth century – commissioned a plan for a new battery at Munden's Point, and two days later he initiated plans for a new fort at Jamestown. He was rewarded for his efforts by the Company in 1708. Some twenty years after Roberts's retirement, an audit by

---

military populations. Both the artillery and infantry regiments were formally disbanded in 1836, after the Company relinquished St Helena to the Crown (Kitching 1947, 4–8).

[25] From the 1670s until the early 1800s, all Company ships calling at St Helena were obliged to deliver one barrel of gunpowder towards the island's defence.

[26] Brooke 1808, 82 note.

[27] The success of this French operation is attributed to the fact that the commander, M. Desduguieres, had previously visited the island where he and his officers had been allowed to roam freely on 'shooting expeditions' and his ships had been allowed to probe the coastal waters around the island, so they could scarcely have been better informed as to the strengths and deficiencies of the island's defences.

[28] From the point of view of the approaching ships, Denholm (n.d., 2) records that 'a complex arrangement of counter-signals had to be exchanged' with the lookouts at Banks's Battery in order to establish that the island remained in British hands – an indication of the continuing precariousness of the Company's control.

Governor Byfield counted 124 cannon deployed around the island, the majority of them (79) at the Castle in Jamestown.[29]

Roberts also instituted the rebuilding of the 'Great Fort' at Jamestown, throwing up substantial stone-built ramparts topped by embrasures for multiple cannon and protected by a dry ditch. Further bastioned defences were built on the headland of Munden's Point, to the north of Jamestown, offering further protection to the roads in Rupert's Bay. Beatson gives further detail of these in his *Tracts*:

> Rupert's Bay, James's Town and Lemon Valley, are the principal landing places on leeward coast. All these are well fortified by fleur d'eau batteries, provided with furnaces for heating shot, and flanked by cannon placed upon the cliffs far above the reach of ships' guns. Mortars and howitzers for showering grape upon ships' decks, or upon boats attempting to land, also provided. In short, it seems wholly impossible to force a descent [sic] at any of those points. Even admitting that the enemy's troops got on shore and succeeded in carrying the fleur d'eau batteries, they would then be exposed to inevitable destruction, from the tremendous fire from the heights, and above all from the immense quantities of grape (or even stones) that might be thrown (with very small charges of powder) from howitzers and carronades of large calibre placed upon the heights, against which it would be wholly impossible to find shelter.[30]

Further improvements to the defences were wrought by Governor Patton, who introduced new forms of gun-carriage that allowed the cannon to be fired more easily at a depressed angle, 'with a facility and accuracy that has astonished every military character who has lately witnessed the St Helena artillery practice'.[31] It was probably he who established the gun emplacement now known as Patton's Battery, commanding the approaches to Jamestown from the lower slopes of Ladder Hill.

By the 1730s the definitive features of the island's principal defence, the so-called 'James Line' that would span the entrance to the James Valley from the sea, had been established, although it would continue to be refined (and repaired in response to sea-damage) over the next half-century (Figure 13). In the following decade the original fort within the town was replaced by a defensive structure known as the Castle, protected by a 6-foot-thick curtain wall towards the sea. A report of 1777 by Major James Rennell mentions thirty-two guns mounted on the line (excluding a saluting battery), twenty-four of them directed out to sea.[32] Supporting cover was provided by substantial emplacements at Patton's Battery, Munden's Point and Ladder Hill Battery. A century later the defences continued to be developed in depth when the earlier fort on the 2,000-foot Knoll Hill was

---

[29] Gosse 1990, 162.
[30] Beatson 1816, lxxxi–lxxxii.
[31] Gosse 1990, 227–28.
[32] For a full description of the evolution of the Jamestown defences, see Denholm 2006, 3–13. By 1825–26 the James Line mounted forty-three garrison guns, twelve carronades, five mortars and four howitzers.

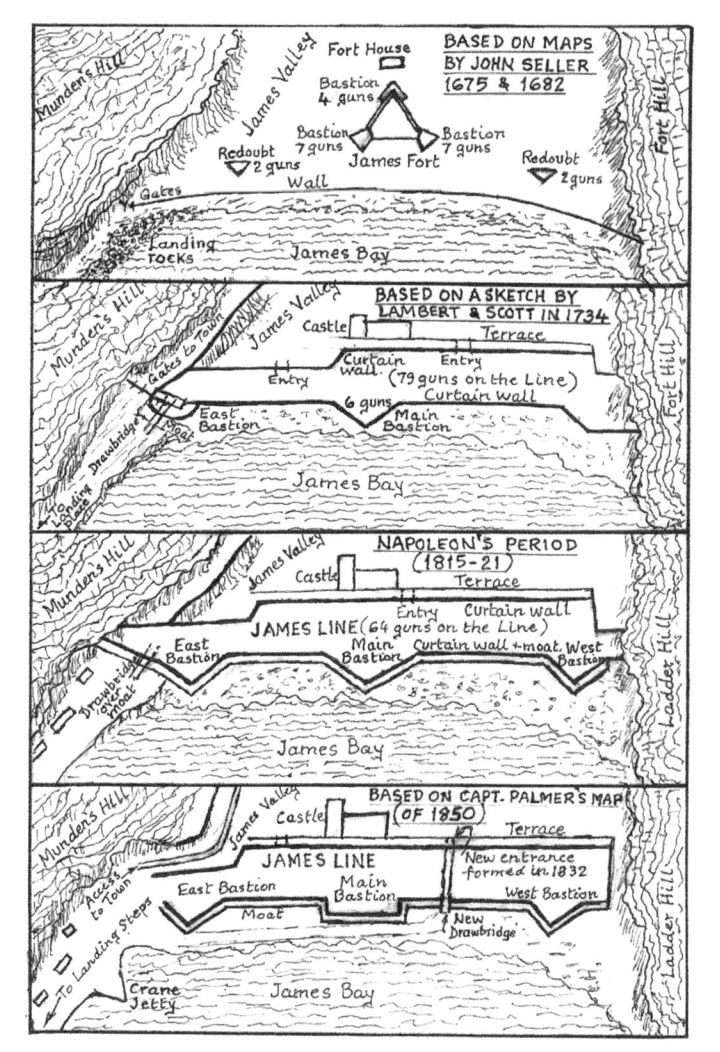

Fig. 13 Plan views showing the four principal building phases of James Fort. From K. Denholm, *South Atlantic Fortress* (2006).

rebuilt and expanded in area by the Royal Engineers (Plate 13). As well as training more guns on the James Valley, the Knoll Hill fort had others facing inland in order to repel an overland attack and a large internal area designed to shelter the local populace in the event of such an attack, though fortunately the measure never had to be implemented. The battery at Upper Munden's evidently was reinforced at this time with 9-inch Armstrong rifled muzzle loaders, with a range of some three miles. Two of these guns, which fired rounds with projecting lugs that engaged with the rifling of the barrel, survive on the battery and two more lie on the rocks below.[33]

[33] https://sainthelenaisland.info/gunsofsthelenant2008.pdf.

Other than at Jamestown, the major fortifications on the island are at Lemon Valley – one of the earliest potential landing places to be defended after the Dutch successfully penetrated there in 1672 – Rupert's Bay (nearly 550 yards in extent) and Sandy Bay (over 400 yards). Banks's Platform, at the north-west extremity of the island, played a major role in monitoring ships heading for Jamestown, which (as mentioned above) were forced by prevailing winds to bear close to the shore – bringing them well within the range of the guns on the platform and those of an additional half-moon battery on the slope above; they would previously have passed through a checkpoint at Flagstaff Bay before turning at Buttermilk Point onto their final bearing for James Bay. These approaches were again heavily defended, as recorded by Lieutenant James Prior, surgeon on the frigate *Nisus*, in 1813: 'Batteries now appear in every direction; guns, gates, embrasures, and soldiers continually meet the eye; so that instead of being, as we might suppose, the abode of peace and seclusion, it looks like a depôt for the instruments of War.'[34] Nearly fifty substantial fortifications, as well as defensive walls built across isolated valleys, are enumerated in Ken Denholm's admirable survey undertaken for the St Helena National Trust.[35]

A benefit was brought to the settlers by the numbers of roads that evidently were constructed by the military to service the more accessible of these forts. They attracted the attention of Charles Darwin in 1836: 'The first circumstance which strikes one, is the number of roads and forts: the labour bestowed on the public works, if one forgets its character as a prison, seems out of all proportion to its extent or value.'[36]

Although remarkably few of these fortifications were ever put to the test,[37] their history throughout the period of their construction and use is one of great complexity. The evolution of strategic planning and of artillery design necessitated periodic amendments to design. Cannon in particular became obsolete with the passage of time and having been manoeuvred into position with enormous expenditure of effort were very often disposed of in the more inaccessible locations simply by being heaved over the ramparts onto the rocks they had

---

[34] Prior 1819, 83.
[35] Denholm 2006.
[36] Darwin 1860, 487. The key role of the military in the upkeep of the roads was made clear as soon as the garrison was withdrawn at the end of the nineteenth century, when drains quickly became blocked and rockfalls remained uncleared (Haliburton 1890, 23).
[37] Lest it should be thought that all was peaceable in the South Atlantic at this time, St Helena continued to be affected by political developments in far-away Europe. One famous incident followed on France's takeover of the United Provinces of the Netherlands to form the Batavian Republic in 1795. When in June of that year news reached the island of the approach of a Dutch merchant fleet, a hastily assembled fleet of East Indiamen led by the *General Goddard*, reinforced by the 3rd rate ship HMS *Sceptre*, succeeded in taking the entire convoy of eight Dutch merchantmen as prize and escorting them into Jamestown.

formerly menaced, where several of them lie today. A more significant threat to the laboriously constructed defences came from the periodic storms that broke over the island: the siting of many of them within the deep ravines that breached the island's natural ramparts made them all too vulnerable to the inundations that periodically swept down from the high interior.[38] Those sited closer to the shoreline were equally in danger from seasonal high seas – not to mention the 'rollers' that periodically break on the island: an early fort at Lemon Valley was lost to one of these in 1667, but the most infamous of them devastated even the massively built defensive line at Jamestown in 1846 (see Plate 16), as well as causing massive damage elsewhere. The combined result of a long history and of the periodic need for renewal is that the structural history of many of the defences is extremely difficult to interpret in detail.

Mention should be made of a number of freestanding round towers built to boost the defences at the turn of the nineteenth century. Although commonly referred to as Martello towers, this identification is challenged by Bill Clements in his account of the 'South Atlantic fortress': those at Prosperous Bay and Thompson's Bay he classifies rather as musketry towers, since they lack any provision for a large-calibre gun on top. Interestingly, the tower at Ladder Hill, built in 1797 and mounting two 12-pounder guns, he characterizes as a forerunner of the true Martello towers that sprang up along the coasts of southern and eastern England some eight years later and subsequently spread through the empire.[39]

As for the progress of the garrison, its condition had been much improved in the 1730s under the command of Captain Cason, with regular drilling of both militia and regular troops. Under later reforms introduced by Governor Brooke at the end of the century, a stretch of wasteland in Jamestown was turned into a formal parade ground – an emblem of the reforms he brought about in discipline and effectiveness. Routine manoeuvres were instituted for the regulars, while the militia too was reformed into a respectable company (Figure 14a). Two companies of black militia were added to the force at this time.[40] All of Brooke's reforms amongst the men and the defences of the island found approval from high-ranking officers from the Indian army passing through on home leave, notably General Sir Archibald Campbell and Lord Cornwallis; only the parsimony of

---

[38] One feature of the reforms introduced to the defences by Governor Brooke in the late 1700s was the resiting of many of them to higher ground, where they commanded the most vulnerable positions from adjacent high ground rather than attempting to block the valleys themselves.

[39] Clements 2007, 83. The same author provides many details of the armaments deployed in all the island's defences.

[40] A more unexpected addition to the militia at this time was two companies of Malays, willingly recruited by Governor Brooke from amongst some 300 prisoners of war seized from Dutch ships and detained on the island. They are reported to have made admirable soldiers during their two-year incarceration on the island and eventually formed the basis of a Malay regiment that saw service in Bencoolen and Ceylon (Brooke 1808, 291–92).

Fig. 14 Badges of (a, left) the St Helena Local Militia; (b, below) the St Helena Rifles.

the Company in providing adequate funding drew their disapproval.[41] One favourable development at this time was the practice of stationing troops on the island for a period of acclimatization before they proceeded to their postings in India. Further approbation of Brooke's efforts came when he was able to send a trained body of over 400 men in response to a plea from beleaguered British forces at the Cape in 1792.

Momentous changes would overtake the island's defences when it played host to Napoleon Bonaparte from 1815, but islanders would have been concerned had they known of certain plans developed by the emperor a decade earlier. They were outlined in a letter to Admiral Denis Decrès, his Minister for the Navy, written from Mainz in 1804:

> There are three expeditions we must make ... To take St Helena and establish a squadron there for several months. For this, 1200 to 1500 men will be needed. The expedition will send 200 men to the aid of Senegal, retake Gorée, and requisition and burn all British establishments along the African coast ... In this way immense harm can be done to the English in the course of three or four months.[42]

Eight ships and 1,500 men were duly assembled with this enterprise in mind, but at the last moment they were diverted instead to Surinam. When Napoleon did eventually reach the island it was under very different circumstances.

During the interlude of Napoleon's incarceration, St Helena accommodated by far the largest numbers of troops it had ever seen, all dedicated to the security of the former emperor, with two regular army battalions totalling some 1,300 men billeted under canvas for the duration of his confinement.[43] These are discussed in greater detail in Chapter 7. An intriguing lost monument of this era took the form of a 'gigantic model' of the island and its defences, constructed to a scale of 1 foot

---

[41] Charles Darwin was struck by the comprehensive nature of the earlier defences, although manning levels had seriously declined by the time he called at the island in 1836: 'It is quite extraordinary, the scrupulous degree to which the coast must formerly have been guarded. There are alarm houses, alarm guns & alarm stations on every peak. I was much struck with the number of forts & picket houses on the line leading down to Prosperous Bay ... a mere goat path ... At the present day two artillerymen are kept there, for what use it is not easy to conjecture ... In some other situations, which were formerly no doubt important, a couple of invalids are stationed; really the places are sufficient to kill the poor men with ennui & melancholy' (Darwin 1933, 412–13).

[42] Reproduced from Martineau 1971, 156. A second expedition against the island, planned for the following year, was similarly abandoned.

[43] When reading accounts of the discomfort and deprivation endured by the French detainees on the plain at Longwood, it is well to remember that those set to guard them endured the same exposure to wind and weather in tents, summer and winter, and with the very worst of diets. Dysentery was endemic in the ranks; many succumbed to disease.

to 1 mile under his own initiative by Robert Francis Seale. Born on the island in 1791 Seale had been educated in England before returning as a junior writer in the EIC's service. An interest in the topography of the island led him to produce first a chart of the coast and its elevations (revealing many more vulnerable landing places than had previously been acknowledged), while at the same time beginning work on the model. When he showed it in its half-finished form to the governor and council they were shocked at the way it laid bare all the island's defences but at the same time they awarded him £50 towards its completion and reported what had transpired to Company headquarters. The directors of the EIC proved equally ambivalent in their response, expressing strong disapproval of Seale's conduct and at the same time arranging to purchase the completed model for a sum of £500. The completed work (over 10 feet in length), showing 'every fortification, house, road, garden, enclosure and division of land', was duly shipped to East India House in its four constituent parts and was placed by the Company in its military seminary at Addiscombe, where cadets received an army education – including an important grounding in the principles of fortification – before proceeding to India.[44] At the closure of Addiscombe the model was transferred initially to the Artillery Museum at Woolwich, but an attempt by Gosse to relocate it ended with the discovery that it had been '"disintegrated" … apparently the official term for "smashed to pieces"'.[45]

Having peaked at a combined strength of some 700 men, the EIC's own artillery and infantry regiments on the island were disbanded with the end of Company rule in 1834.[46] Along with Major-General George Middlemore, the first governor appointed directly by the British government, there arrived on the island the 91st (Argyllshire) Regiment of the regular army. One of Middlemore's first actions was to disband both the infantry and artillery regiments formed under the Company: 'there was no ceremonial parade or leave-taking; the Colours were cut up and distributed among the officers, and the regiments were just summarily disbanded, as it were under guard, and faded away';[47] it was an unworthy end to a century and a half of service. In 1842 the island was garrisoned by a European regiment raised expressly for the purpose and styled the St Helena Regiment. In a curious

[44] MacGregor 2018.
[45] Gosse 1990, 304–08. Chancellor and Van Wyhe (2009, 523) confirm that the model survived until at least 1930; thereafter the museum's records were destroyed by incendiary bombs in 1940. In a tantalizingly titled article on 'Charles Darwin's St Helena model notebook', Chancellor (1990) transcribed the fragmentary text surviving from Darwin's notes made while he refreshed his memory of the volcanic structure of the island by reference to Seale's model then at Addiscombe, but sadly it reveals no detail whatever about the model itself.
[46] Kitching (1947, 7) records that by the final decades of Company rule the officers of the St Helena regiments were highly professional, being frequently the sons of landowners on the island who had completed their education in Britain and had graduated from the Addiscombe military seminary.
[47] Kitching 1947, 8.

throwback to seventeenth-century practice, inducements to join the regiment initially included the provision of land for cultivation, but the practice was quickly abandoned and within fifteen years, amid growing discontent and with a declining reputation, the regiment was abandoned.[48] The militia, meanwhile, had been re-formed nine years earlier, equipped with new uniforms provided at government expense on the pattern of the Rifle Brigade.[49]

The progressive attrition experienced by the garrison thereafter would have serious consequences for the island's economy. A major reappraisal of its strategic significance in naval terms contributed to and hastened this process. Judged at one time 'the key to the whole South Atlantic',[50] in the age of steam St Helena found itself once more on the road to nowhere: there was simply no need for anyone to go there. For a time the Navy considered maintaining it as a coaling depot, but quickly decided that such a move would provide only limited advantage, would contribute to making the island a target for any enemy, and would require the deployment of further defensive forces; when the War Office balked at the idea of deploying troops merely to defend the Navy's coal depots, insisting that any such role should be no concern of theirs, the proposal sank without trace.[51] Finally, and despite a spirited campaign of lobbying in favour of maintaining the strength of 'the citadel of the South Atlantic',[52] the early years of the twentieth century saw the garrison first reduced by half and then withdrawn altogether. At the same time, the St Helena Volunteers were disbanded, leaving the island for the first time since the seventeenth century without a military presence. The perceptive question posed by William Webster (as quoted at the head of this chapter), namely 'of what importance or value could be such a rock as this?' when Britain had lost the sovereignty of the seas, appeared to have received an eloquent answer from the Lords of the Admiralty.

Two World Wars would largely pass St Helena by, although their history too is marked on the face of the island.[53] A battery for two modern 6-inch breech-loading guns, supported by underground magazines, had been constructed at Ladder Hill in 1903, just five years before the final withdrawal of the garrison; abandoned by the Army, the guns were reactivated in 1912 after responsibility for the island's defence was transferred to the Royal Navy and the Royal Marines. In 1938 a decision was taken that the guns should be shipped back to Britain,[54] but with the outbreak of World War II they gained a reprieve and responsibility for them was transferred to a newly formed unit, the St Helena Coast Battery of the

---

[48] Kitching 1937, 53.
[49] Chartrand 2011.
[50] Mellis 1875, 44.
[51] Ekoko 1984; Gray 2018, 26–29.
[52] Haliburton 1890.
[53] See Clements 2007.
[54] Much of the earlier artillery was indeed shipped back to England at this time as scrap, to be melted down for the war effort.

Royal Artillery.⁵⁵ The Elswick Mark VII coastal defence guns at the Ladder Hill bastion (Plate 14) – by now the island's key defensive site – had an impressive range of seven miles, but spent the entire war without being fired in anger. With the conclusion of hostilities, their breechblocks were removed and dumped at sea; at the same time, the St Helena Coast Battery was disbanded.

Infantry regiments, dependent on sheer manpower rather than massive installations, tend to leave fewer traces of their coming and going. At the opening of the twentieth century the St Helena Volunteer Rifles (Sharpshooters) formed such a unit; founded initially to assist in guarding and providing for the prisoners of war held on the island from the Boer War (Chapter 8),⁵⁶ they later saw service in World War I: a monument to one of their riflemen, who died on 4 May 1915, survives in the cemetery of the cathedral on the island. With the outbreak of World War II, the St Helena Rifles (Figure 14b), formed initially in 1939 as a local defence force, was quickly merged with the regular army. Troops were shipped to Britain and remained in service until the regiment was disbanded in 1946, when many men transferred to other regiments as regular soldiers.

At sea, the war in far-away Europe impacted directly on the island on only one occasion, with the sinking by a German submarine of the *Darkdale*, a tanker of the Royal Fleet Auxiliary. The *Darkdale* had arrived on station in James Bay in August 1941 and in the following months successfully refuelled a number of Royal Navy warships, but on 22 October she was struck while at anchor by four torpedoes from the *U-68*, with the loss of forty-one men.⁵⁷ The *Darkdale* was riven in two by the force of the explosions and sank on the spot. A continuing reminder of her presence persisted in the form of an oil slick that seeped from her ruptured tanks over the next seventy-five years, posing a significant environmental threat to the island, until in 2015 the tanks were drained and the ship's ammunition was recovered by divers.

---

55 Clements (2007, 87–88) mentions the establishment of a second 6-inch gun emplacement at Munden's Hill, the guns from which, as a result of a strategic rethink in 1940, were repatriated to reinforce coastal defences in post-Dunkirk Britain. Other authorities deny that such guns were ever installed, although a World War II searchlight station was created there, operated by the St Helena Rifles.

56 Heunis (2019, 15) mentions that guard duties were provided at various times by detachments from the Royal North Gloucestershire, the Berkshire and the Wiltshire Regiments.

57 On the same long-range patrol from her base at Lorient in Brittany, the *U-68* had already sunk one ship off the Canary Islands and had escaped damage in an action at the Cape Verde Islands; after sinking the *Darkdale* she would find further victims in the *Hazelside*, destroyed some 690 miles south-east of St Helena, and the *Bradford City*, off South West Africa (present-day Namibia), before returning damaged to her base. https://en.wikipedia.org/wiki/German_submarine_U-68_(1940).

'The citadel of the South Atlantic'

Signals and telegraph

By the turn of the nineteenth century the system of communication on which the defensive stations depended was brought to a new level of effectiveness by Governor Patton. The positioning of Jamestown, on the north-westerly leeward coast, rendered the efficient working of the system imperative, since hostile ships were most likely to emerge under full sail on the diametrically opposite quarter of the island and the deployment of appropriate defensive forces depended on alarms being relayed by lookouts posted strategically across the intervening five miles or more of mountainous interior. Evidently Governor Brooke had devised a system of communication only a decade or so earlier, but this was now deemed inadequate. Patton took the trouble to describe the working of his system, which he judged operated 'in a simple, easy, clear, and distinct manner without a possibility of error':

> I was led to prefer the sort of Telegraph I have adopted, composed of a frame of wood and balls, because our stations are generally upon heights having the sky behind which renders such objects distinctly visible ... With only four balls, one hundred different signals can be distinctly made, in the numerary manner, in that the system is a species of communication in cypher – a ball hoisted above the frame indicates a hundred, the progress in this manner being unlimited.[58]

From 1806 an army sergeant was appointed Superintendent of Telegraphs and Maritime Signals. Further signals were made with the aid of flags and the firing of alarm guns: the latter were arranged to relay signals from Prosperous Bay signal station to Jamestown, via the intermediate station at Alarm House and others elsewhere.

Patton's new system – designed by himself and produced at little cost – covered every quarter of the island so effectively that (according to one of his successors) no vessel could approach within sixty miles of the island without being detected. Royal Navy and EIC ships could all be identified long before they reached Jamestown. In this way the telegraph system had, in Governor Beatson's words,

> placed the whole island under the eye of the Governor; for he is instantly apprised of every material occurrence in any part ... and, with equal celerity, he can convey his orders wherever they may be necessary, both during the day and night ... With such means of receiving information, and of sending orders, a Governor of St Helena is as fully prepared to oppose a vigorous resistance, at every point in his extended line of defence, as if he commanded within a small fortress.[59]

---

[58] Quoted (with other details here) from the chapter 'St Helena's pioneer telegraph system' in Hearl 2013, 77–87. Lanterns replaced the balls during hours of darkness.
[59] Beatson 1816, lxxxvi.

A flavour of life on the island fortress is conveyed by the journal of William Burchell (Chapter 6), maintained throughout Burchell's stay from 1805 to 1810. In it are repeated allusions to 'alarms' being announced by cannon fire at the first sighting of sails on the horizon: typical is an entry for 17 April 1807, when, 'At about 9 this morning, a signal was fired for a fleet being in sight, and as usual the whole Island got under arms.' The system evidently worked perfectly well and was widely understood: at least, that is the implication of the chaos that followed the imposition of a new system under the governorship of Colonel Lane, which seems to have been constantly misread and misinterpreted. On 12 December 1807, for instance, 'This morning an alarm was fired for a fleet of 10 ships and all the Garrison were under arms, but it proved a false alarm, for not one ship was in sight. This is the effect of changing the signals.'[60]

In his key article on the island's signalling system, Trevor Hearl quotes Captain Charles Dixon of the Royal Engineers as reporting to London c.1815 that there were at that time fourteen signal posts, 'all with alarm guns, generally long 9-pounders': evidently neither Patton's nor Lane's system had stood the test of time – quite possibly on account of the propensity of one or other of the visual signal stations to be shrouded in cloud cover at critical moments, although signal posts and flags evidently formed adjuncts to the guns, providing visible confirmation of the alarms fired. Hearl lists the positions of the known posts, some of which are recorded in a topographical view by Bellasis (Plate 15), and he reproduces images of several of the stations; he also confirms that the entire system was abandoned for military purposes at the time of the island's reversion to the Crown in 1834.[61] The system of 'Time-Ball Office and Telegraph Stations' introduced by the colonial administration three years later survived until the withdrawal of the island's garrison. An observer in 1890 witnessed the following poignant scene:[62]

> A cart-load of flags, seen yesterday on its way down from High Knoll to Jamestown, was a significant sight, for the time balls, and the flags and signals from the flagstaff will henceforth cease to announce noon, the Queen's birthday, the arrival of steamers and other important events.
>
> The flagstaff at High Knoll signal station has been used for upwards of fifty years for the purpose of giving the time, and (about 25 years) signals denoting the arrivals of steamers and men-of-war. This boon will be greatly missed by the people in the Country, as most of them ... depended solely on it for time and information.

---

[60] Burchell's journal, 8 and 12 December 1807: see Castell 2011.
[61] Kitching (1947, note 5) mentions the adoption thereafter by the Army of 'a contraption like a cricket scoring-board with numbers'.
[62] Haliburton 1890, 23.

'The citadel of the South Atlantic'

By 1864 the commanding officer of the Royal Engineers could be found drawing up estimates for the introduction of an electric telegraph system to link the various lookout points with Jamestown, although the visual signalling system survived alongside it for a time.[63] Within a few years the telegraph system was available for sending private telegrams within the island; by 1886 the telephone was introduced – remarkably within ten years of its invention. Within another decade the arrival of the submarine cable carrying signals all the way to England would bring the island's officialdom into unprecedentedly direct contact with the outside world, but it would take another century and more for the population at large to gain equal access to intelligence of the world as it impacted on the island.

---

[63] Denholm n.d., 10–11. Denholm notes that electrical telegraph communications had been used in the Crimean War during the previous decade.

Plate 1 *St Helena, taken from the Sea*, from George Hutchins Bellasis, *Views of St Helena* (1815). The island is pictured from the north-west, 'about eight miles from the land', with James Bay on the left; full force is given to the impregnable aspect it presents to mariners.

Plate 2 Plan of St Helena by James Imray & Son (1803–70). Imray's well-known series of blue-backed maritime charts frequently relied for their detail (as here) on earlier published maps, assembled for the benefit of mariners.

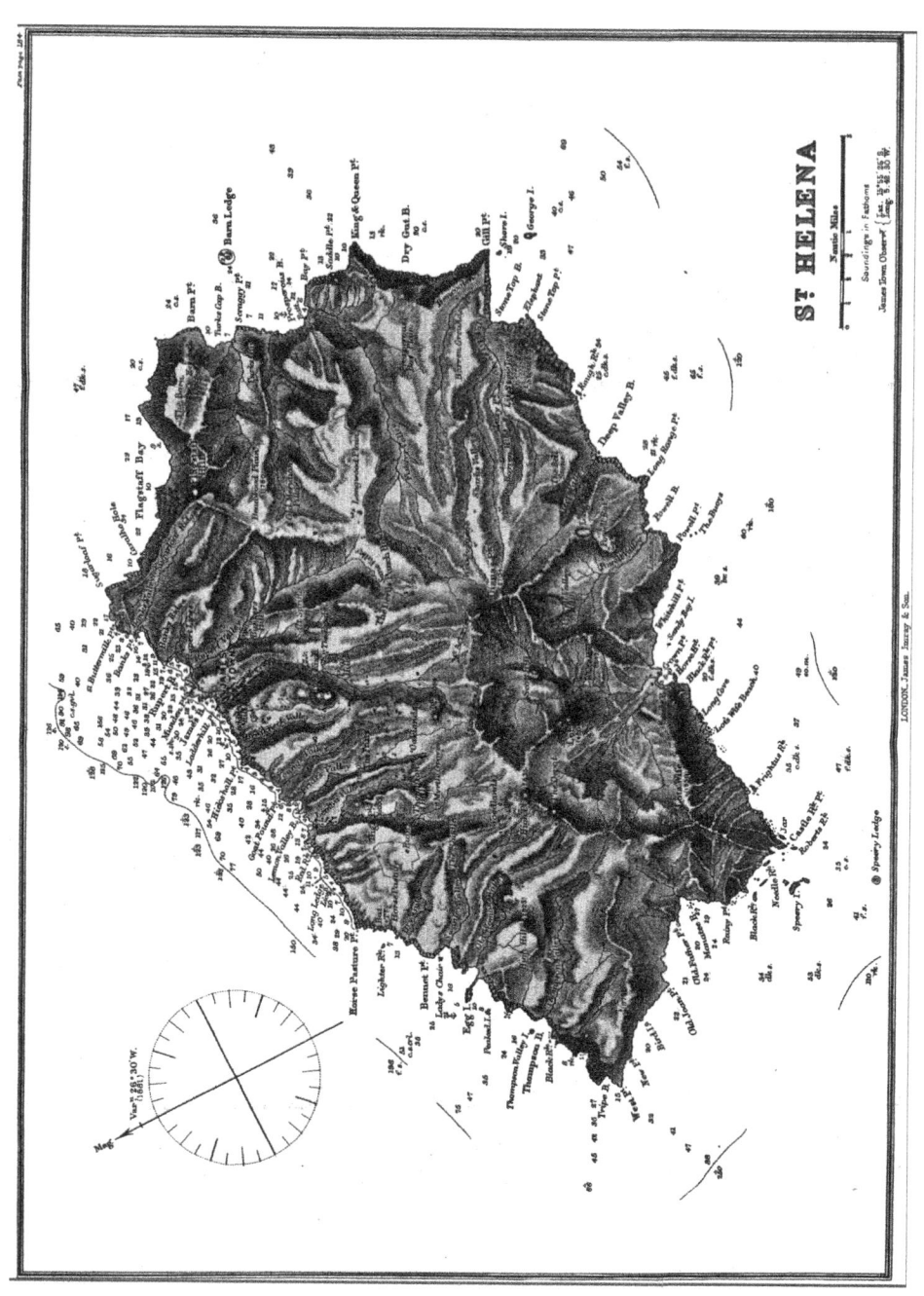

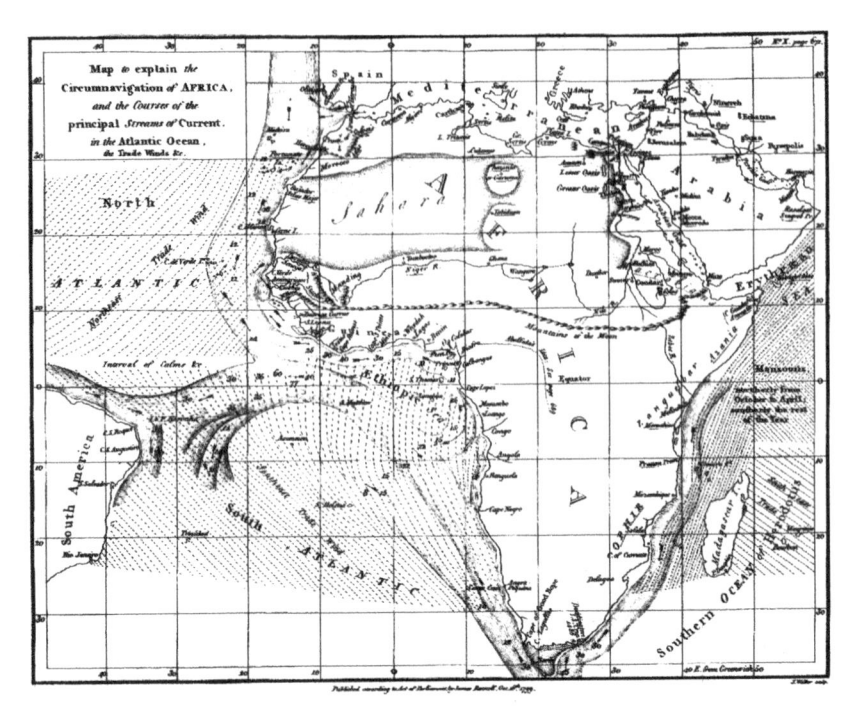

Plate 3   Chart by James Rennell, including 'the Courses of the principal Streams of Current in the Atlantic Ocean, the Trade Winds, &c.' Compiled on Rennel's homeward voyage from India in 1777 and published in 1799.

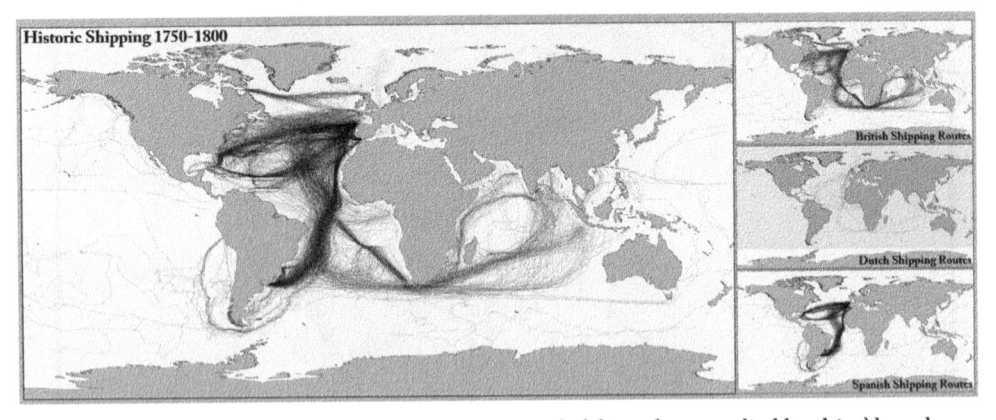

Plate 4   Historic shipping routes, 1750–1800, compiled from data supplied by ships' logs, by Professor James Cheshire. The importance to British vessels of the homeward-bound course via St Helena is immediately apparent.

Plate 5   A vestige of the original cloud forest cover now forming the Diana's Peak National Park.

Plate 6   View of the eroded landscape with storm watercourses that characterizes much of the coastal fringe of St Helena.

Plate 7 A few samples of the cargo of porcelain salvaged from the wreck of the *Witte Leeuw*, lost in 1613 following an explosion in the powder magazine.

Plate 8 Proclamation claiming St Helena for the Dutch Republic, 15 April 1633, signed by Jacques Specx, former Governor-General of the Dutch East Indies (1629–32).

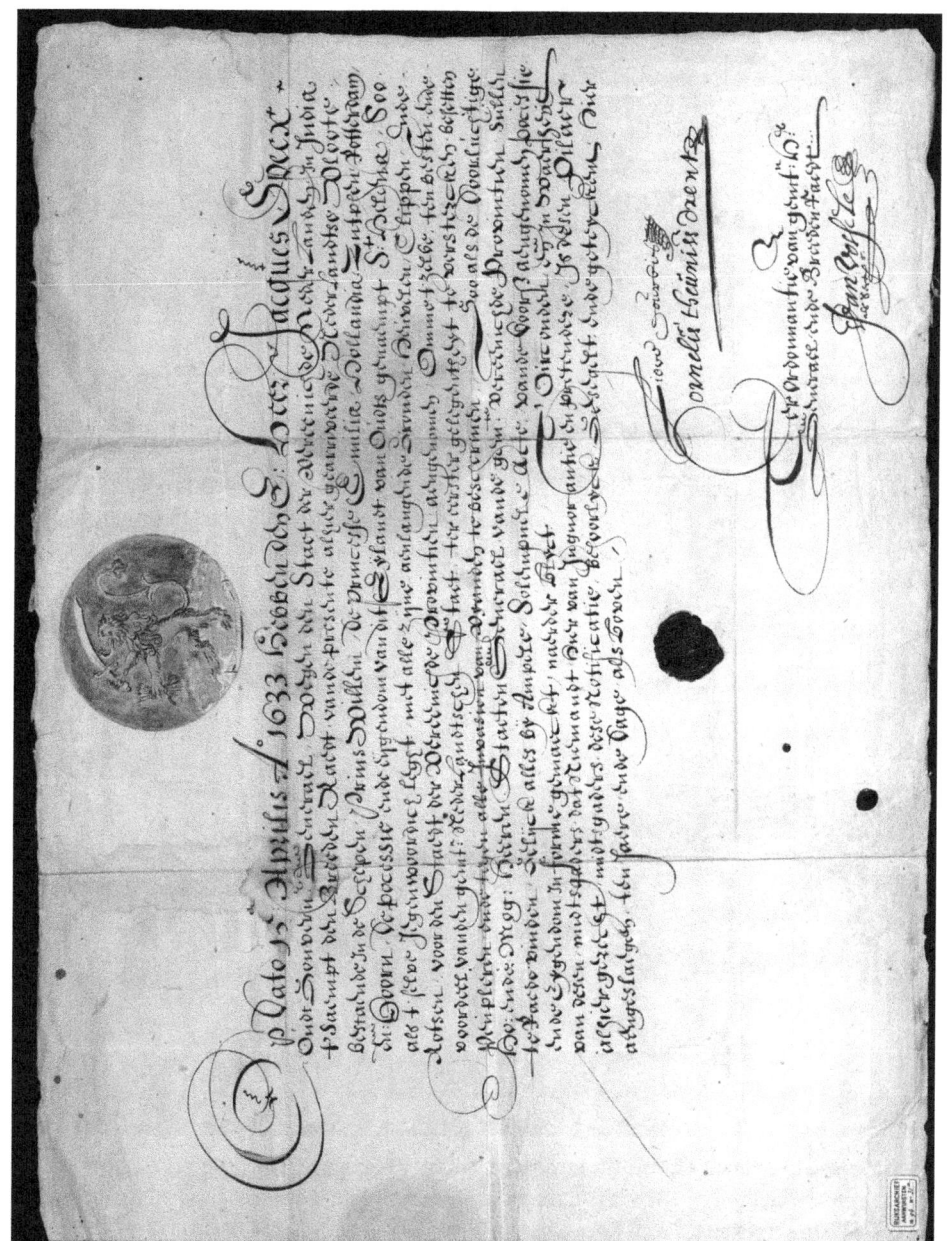

Plate 9   The St Helena Wirebird (*Charadrius sanctaehelena*), drawn by Mrs J. C. Mellis for reproduction in J. C. Mellis, *St. Helena: a physical, historical, and topographical description of the island* (1875).

Plate 10   Plantation House, the governor's country residence. The property is shown in 1819 when the extent of the nurseries and experimental gardens reached their maximum extent. From Kerr, *Series of Views in the Island of St Helena* (1822).

Plate 11 New Zealand flax, once the basis of a thriving industry but now one of the most invasive plants on the island posing a threat dramatically conveyed in this photograph of the present-day clearance campaign.

Plate 12 Share certificate, issued 23 December 1837 – the year the St Helena Whale Fishery Company was formed. When the traditional whaling grounds off the New England coast were lost to the British following American independence, the focus moved to the South Atlantic.

Plate 13 Knoll Hill fort, the most highly developed (and one of the most inaccessible) fortifications on the island. Its long architectural history was brought to completion by the Royal Engineers in 1874.

Plate 14 The Ladder Hill battery, mounted with an Elswick Mark VII coastal defence gun, never fired in anger.

Plate 15 View of the Jamestown approaches with multiple signal stations on the heights and in the valley. From George Hutchins Bellasis, *Views of St Helena* (1815).

Plate 16 *The rollers of 17 February 1846, taken from the Harbour Master's office*, '… in which day Thirteen Vessels (mostly captured Slavers) were wrecked … as well as Private Property to the amount of £10,000 destroyed – it. Was also remarkable that the agitation of the Water was confined to about 500 yards from the shore, beyond which distance the sea was perfectly calm.' Coloured lithograph by T. Picken after Lieut. F. R. Stack.

Plate 17   Edmond Halley's star chart of the southern celestial hemisphere, engraved in 1678 by James Clerk, dedicated to King Charles II.

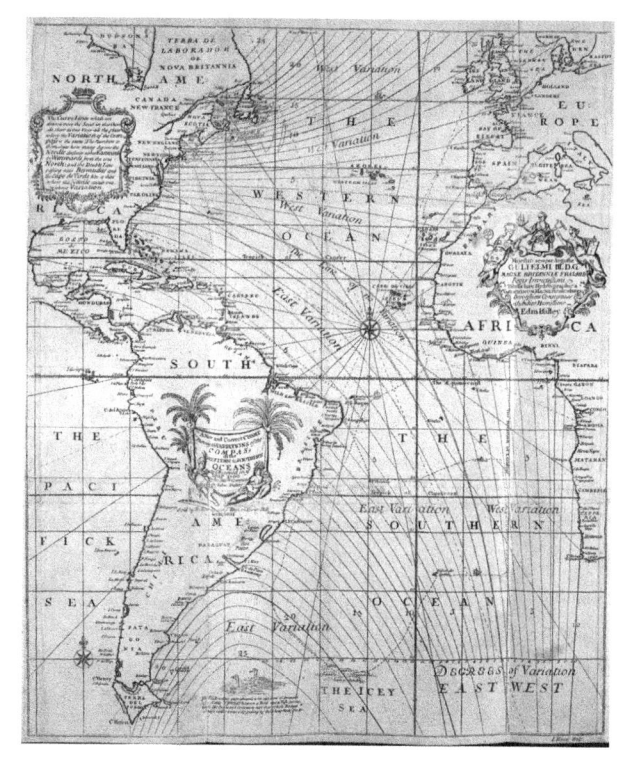

Plate 18   Edmond Halley's *New and Correct Chart shewing the Variations of the Compass in the Western and Southern Oceans* (1700). Halley was the first to chart magnetic variation in the form of isogonic lines in the ocean.

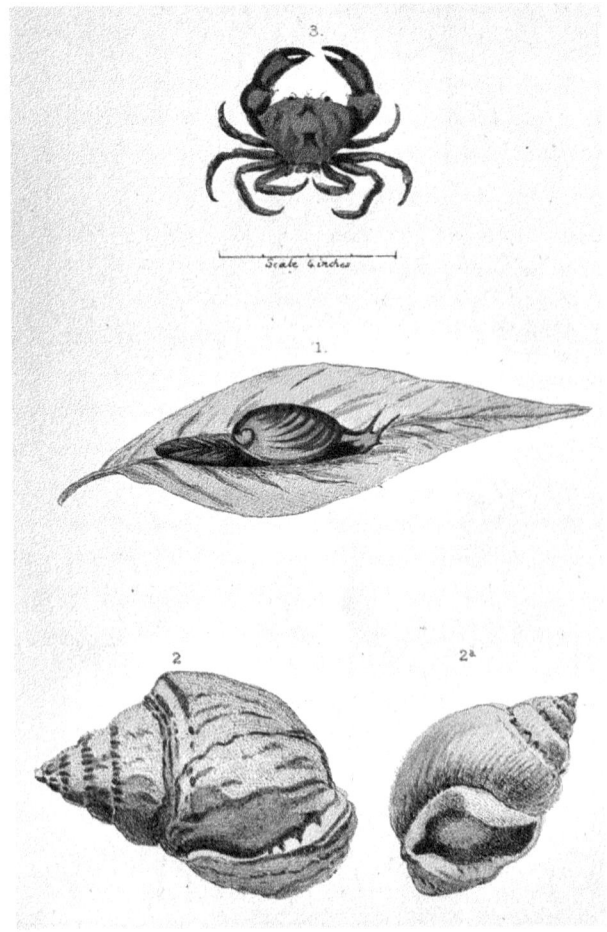

Plate 19 (above) William Burchell at the summit of Sugar Loaf, with Diana's Ridge in the background, 8 December 1807.

Plate 20 Two illustrations from John Charles Mellis's *St Helena: A physical, historical, and topographical description of the island* (1875); (a, left) pl. 22 (variety of specimens); (b, opposite) pl. 54 (*Dickinsonia arborescens*).

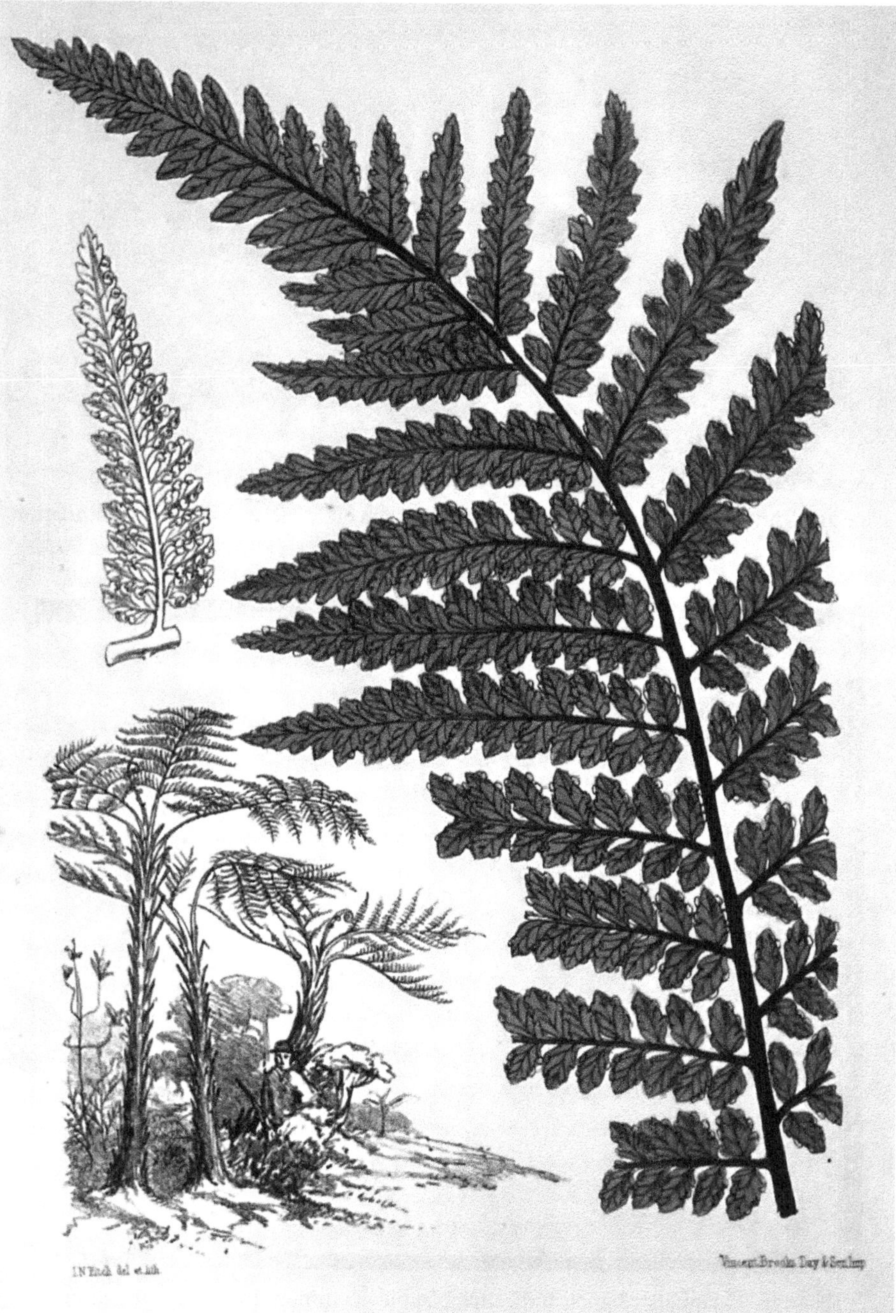
J.N Fitch del et lith
Vincent Brooks Day & Son Imp

Plate 21   The Briars, site of Napoleon's initial quarters on St Helena, before his translation to Longwood House.

Plate 22   Manual of signals indicating Napoleon's minute-by-minute situation, drawn up for Governor Wilks, 1815–16.

Plate 23   Longwood House following its expansion to accommodate its Napoleonic occupants, anonymous view and ground plan.

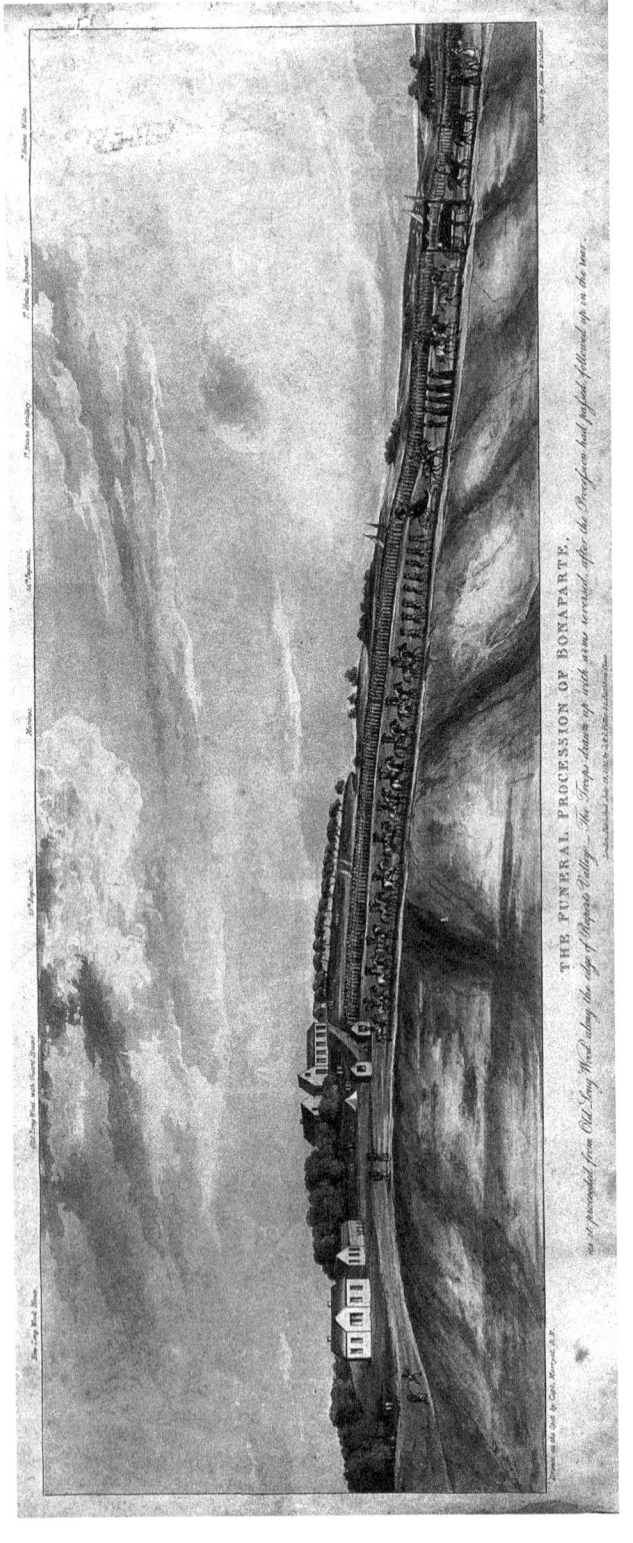

Plate 24   Napoleon's funeral. (a, above) The cortege sets out from Longwood: aquatint by Alken and Sutherland after a drawing by Captain Frederick Marryat; (b, below) it approaches the site of the tomb, from Kerr, *Series of Views in the Island of St Helena*), 1821.

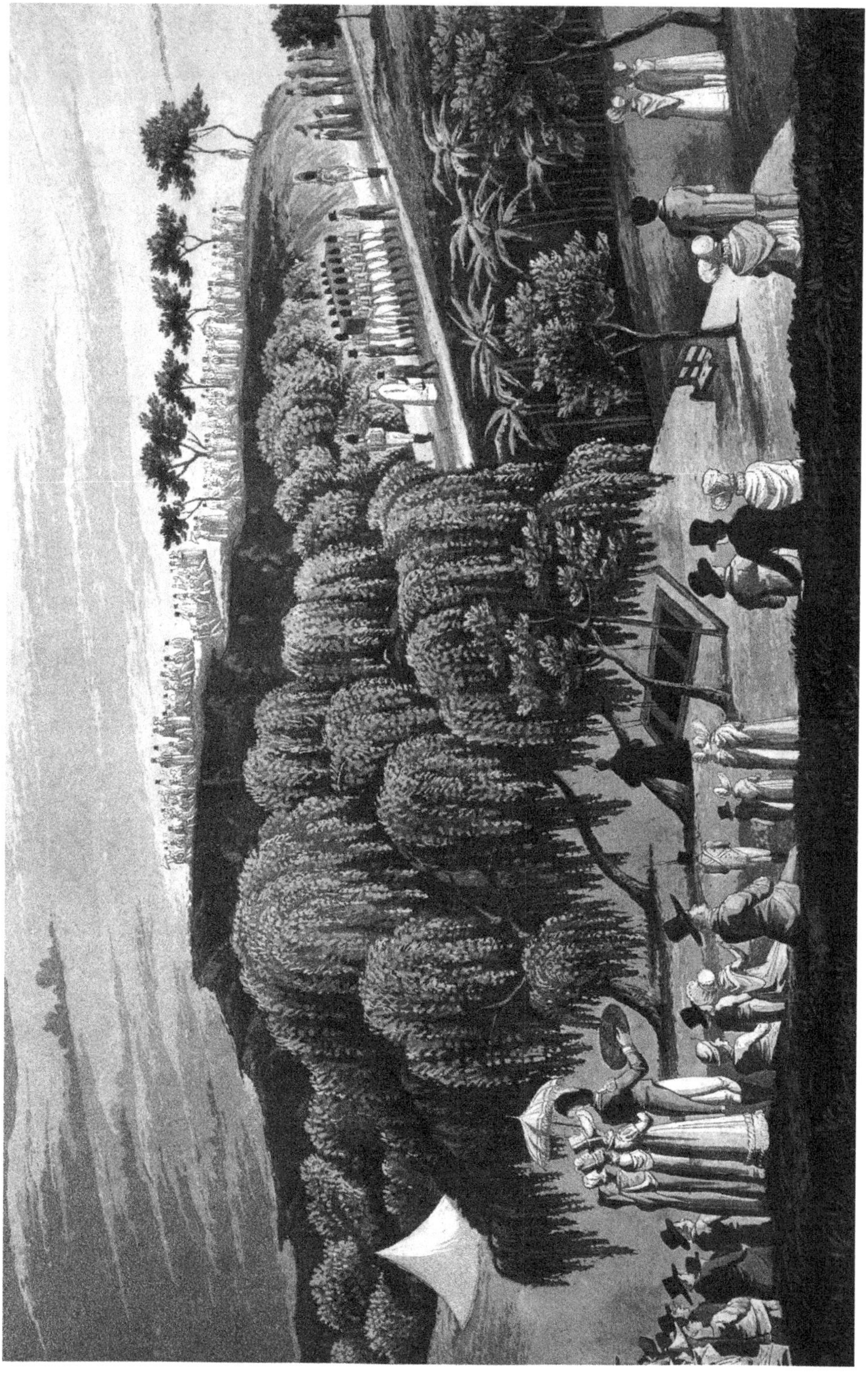

Plate 25   Erich Mayer, *Camp life*, watercolour. Tents functioned not only as sleeping and eating quarters but also as workshops for craft activities. Mayer, a German national, had been captured at Elandlaagte, near Mafeking; he died on St Helena and is buried there in the prisoner-of-war cemetery.

Plate 26   Replica of a chessboard made by a Boer prisoner of war on St Helena.

Plate 27 RMS *St Helena* (in service 1978–90), following her refit in 1978.

Plate 28 RMS *St Helena* (in service 1990–2018), at anchor in James Bay.

Plate 29   The pilot of a Boeing 757-200 – capable of carrying up to 200 passengers and since 2020 the largest aircraft so far to have landed at St Helena – takes no chances with an assertive touchdown, despite the limp windsock by the runway.

Plate 30   An established area of the endemic Gumwood *Commidendrum robustum*, at the Millennium Forest project (once site of the Great Wood), April 2023.

6
Scientists in transit:
St Helena as a site for scientific investigation

For over three centuries, while the attention of the government and population of the island was focused on matters of survival and improvement, St Helena played host to a succession of scientists pursuing research in a variety of fields, attracted by the island's isolated geographical location in the South Atlantic, the geological history locked in its rocks, the indigenous flora and insect fauna with which it was blessed (and subsequently robbed), or the denizens of its surrounding seas. Few places on the planet can have proved so popular a laboratory for the compiling and testing of scientific theories and for field surveys – a practice that survives today; recent years have seen the island used as a base for exercises as diverse as satellite tracking and deep-sea oceanography.

A constellation of astronomers

During the development of (on the one hand) modern astronomical practice in Britain from the seventeenth to the nineteenth century and (on the other) the extension of the British Empire into the southern hemisphere and the Far East, St Helena proved repeatedly inviting to aspirant astronomical observers. Despite its strategic location on the surface of the globe, however, the island's topography and the vagaries of its weather conspired on more than one occasion to frustrate the ambitions of would-be observers, however carefully-planned their expeditions.

EDMOND HALLEY

Edmond Halley (1656–1742) arrived on the island in 1676, having just turned twenty-one but already with a respectable reputation as an astronomer. He had begun making observations while still at school and continued to do so at Oxford, publishing his results in the *Philosophical Transactions* of the Royal Society. It is from his correspondence with Henry Oldenburg, the society's secretary, that we first learn of Halley's ambitions to observe the southern stars and of his nervousness that a rival astronomer then preparing a book in Paris might pre-empt his aim to be the first observer in this field:

> ... if that work be yet undone, I have some thoughts to undertake it my self, and go to St Helena ... by the next East Indie fleet, and to carry with me, large and accurate Instruments, sufficient to make a good catalogue of those starrs, and to compleat the Celestiall globe ... I will willingly adventure myself, upon this enterprize, if I find the proposition acceptable, and that the East Indie company will cause me to be kindly used there, which is all I desire as to my self, and if I can have any consideracon for one to assist me.[1]

Amongst the supporters of Halley's proposal was John Flamsteed, the first Astronomer Royal, whom he had 'carefully assisted' on one or two occasions. Flamsteed was already preparing the instruments at Greenwich that ultimately would facilitate his own compilation of an accurate catalogue of northern stars with the aid of the telescope; his aim was to improve the precision of the survey compiled by the Danish astronomer Tycho Brahe at the end of the previous century – before the invention of the telescope. The project fired Halley's imagination, prompting him to leave Oxford without taking a degree but promising himself that he would now 'do something acceptable to the learned world'.[2] Having duly assembled support from a number of other backers (including King Charles II, who requested the EIC to provide Halley and a companion with a free passage),[3] he sailed in November 1676 on the ship *Unity*, with the intention of undertaking a corresponding survey for the southern hemisphere from St Helena – at that time the southernmost point on the globe under British rule. An additional ambition – one that had dictated the need for his immediate departure – was to observe the transit of Mercury across the sun's disk on 7 November 1677 – an event which Halley realized would provide an important opportunity for determining the distance of the sun from the Earth.

The instruments with which Halley arrived on the island were amongst the most advanced of their kind, and he was careful to describe them at length:

> When my voyage had been assured I had a sextant made, of which the radius was five and a half feet in London measure: the framework is of iron, the limb and the scales are of brass, it is fitted with telescopes, and so that it can be rotated easily and precisely in the necessary operations; it is placed on two toothed semicircles which intersect at rightangles, and which rotate on an endless screw, adjusting the plane of the instrument in such a way that without any labour and the greatest readiness, they adjust themselves to any situation for any two stars

[1] Oldenburg 1986, 368–69.
[2] Halley to Oldenburg, 8 August 1676: Oldenburg 1986, XIII, 27.
[3] Much of the finance was supplied by Halley's father, a wealthy merchant. The sea passage was requested 'in order to the transportation of themselves and their necessaries for St Helena' (quoted in Jones 1957, 176–77). Halley's companion is now identified as James Clerk. Tatham and Harwood (1974) suggest he may be the same person who later engraved Halley's planisphere. Although Halley wrote that in Clerk's company 'all the good fortune I have had this voyage consisteth', Flamsteed laid the blame for many of the expedition's shortcomings at Clerk's door (Cook 1998, 77–78).

at the same time. I have also a quadrant which I have used before, of which the radius is about two feet; I have used it infrequently in celestial observations, but only to find the elevations of the Sun which are necessary to correct the faults and inequalities which usually occur in the timekeeping of clocks. Lastly I prepared several telescopes of various lengths, of which the longest is twentyfour feet, with two micrometers to measure the arc between the object and the bearing of the telescopes.[4]

It has been observed that the evidently magnificent sextant was a version (only slightly scaled-down) of the first major instrument installed at the Royal Observatory at Greenwich by Flamsteed;[5] its advanced design (Figure 15) contrasts with the rather modest scale of the observing station established for the benefit of the astronomers on the north-facing slope of the ridge today known as Halley's Mount, in the centre of the island.[6]

Contemporary records are silent about the construction of the observatory,[7] but it is now tentatively identified with the remains of a two-roomed stone-built structure (the smaller of the two chambers having been added when it was later adapted as a signal station), oriented true north and south (Figure 16). It would have had, perhaps, a light roof that could be opened up to allow for observations, while the walls shielded the unwieldy instruments from the wind. Although he could scarcely have been more assiduous in his observations – reportedly never going to bed when a clear night sky presented itself – the weather was unkind to Halley for lengthy periods of time, severely limiting his opportunities. He tells of his frustrations in a letter to one of his patrons, Sir Jonas Moore, Surveyor-General of the Ordnance, dated 22 November 1677:

> I hoped still that we might have some clear weather when the Sun came near our Zenith, that so I might give you an account that I had near hand finished the Catalogue of the Southern Stars, which is my principal concern; but such hath been my ill fortune, that the Horizon of this Island is almost always covered with a Cloud, which sometimes for some weeks together hath hid the Stars from us, and when it is clear, is of so small continuance, that we cannot take any number of Observations at once; so that now, when I expected to be returning, I have not

---

[4] The instruments are carefully itemized in Halley's *Catalogus Stellarum Australium* (1679), in a preface dedicated to 'Astronomiæ Studiosis'; the translation quoted here is from Tatham and Harwood 1974, 492.

[5] Alan Cook observes that Flamsteed's sextant was certainly made by artificers in the Ordnance Office, under the control of Sir Jonas Moore, and that the same technicians 'probably made the quadrant and sextant that Halley took to St Helena' (Cook 1998, 38).

[6] The name is first attested in 1682, only five years after Halley's visit to the island (Tatham and Harwood 1974, 491). For images of the site see https://academic.oup.com/astrogeo/article/54/4/4.7/182079.

[7] The 'Consultations' or minutes of the Governor's Council are extant only from 1678 onwards.

Fig. 15 Flamsteed's 7-foot sextant, on which Edmond Halley's instrument, brought to St Helena, was based. The successful installation of such a cumbersome yet sensitive instrument in Halley's primitive observatory is scarcely less remarkable than its successful operation. From Flamsteed's *Historiæ Coelestis Britannica*, vol. III (1725).

Scientists in transit: St Helena as a site for scientific investigation

Fig. 16 Halley's observatory. The location had been lost until 1968, when the footings of the stone building were rediscovered. The plaque within reads 'The site of the observatory of Edmond Halley. He came to catalogue the stars of the southern hemisphere 1677–1678'.

finished above half my intended work; and almost despair to accomplish what you ought to expect from me. I will yet try two or three months more, and if it continue in the same constitution, I shall then, I hope be excusable if in that time I cannot make an end. However it will be a great grief to be so far frustrated in my first undertaking.[8]

According to Robert Hooke, the 'abundance of mists and moysture ... unglued the Tubes' of Halley's telescopes,[9] while his notes and charts were always critically at risk from damp. Halley himself adds further that 'the mighty winds, and extraordinary swift motion of the Clouds hindred the exactness of the Observations'.[10]

Nonetheless, Halley did succeed in establishing the positions of several hundred stars and in making the first complete observation of the Transit of Mercury, which provided him with the data he sought for calculating the

---

[8] The text of the letter is printed as an appendix to Robert Hooke's *Cometa, or, Remarks about Comets* (1687), 75–76.

[9] Robert Hooke, *Of Spring*, quoted in Chapman 1994, note 33.

[10] Halley to Jonas Moore, 22 November 1677, in Halley 1932, 39–41. In their account of excavations on the likely site of the observatory, Tatham and Harwood (1974, 500) noted its positioning behind a sheltering knoll, beyond which 'the trade wind blows at ten knots and more – enough to upset delicate observations'.

distance of the sun from the Earth (a topic to which he returned later in his career). At the same time, he realized that the Transit of Venus (not scheduled until 1761 and again in 1769) would provide opportunities for even more accurate calculation of this fundamental measurement in astronomy, known today as the astronomical unit. His assessments in this respect were duly adopted by the scientific establishment and played an important role in ensuring that further expeditions to the southern hemisphere would be mounted to record those events (see below).[11]

Halley left St Helena on the *Golden Fleece* at the beginning of March 1678 and the following year brought the results of his observations to publication with the title *Catalogus Stellarum Australium* (1679), translated shortly thereafter into French. The swiftness of its appearance, and that of a 'planisphere and description of the stars of the southern hemisphere' which he exhibited to acclaim before the Royal Society on 7 November 1678 (Plate 17), suggests that a good deal of the preparatory work must have been carried out in the course of the voyage home. Halley's achievement in charting during the course of his twelve-month stay, under less-than-ideal conditions, the celestial latitudes and longitudes of 341 stars not recorded in Brahe's catalogue seems little short of miraculous.[12] In time he would succeed his mentor Flamsteed (who styled Halley appropriately 'our southern Tycho') as the second Astronomer Royal, although relations between the two men had become embittered in the meantime.

In 1686 Halley published the second report on his expedition to St Helena, in the form of a learned paper and a chart on the subject of trade winds and monsoons. Solar heating is identified here as the primary cause of motions in the atmosphere. He also established the relationship between barometric pressure and height above

11   Halley's attempts to accurately determine the longitude of St Helena at this time would serve him in the future when he returned on board the *Paramore*: 'by my account of the Ships way this Island Should be 8°:40' more Westerly than London, which is two degrees more than by Celestiall observation I have found formerly, and so much I conclude I have been Sett to the Eastwards by Some Current since ... I left the Isles of Tristan da Cunha' (Thrower 1981, 179–80).

12   Brahe's work in locating the northern stars provided important overlapping reference points by which Halley's observations could be tied to the more familiar star map of the northern skies. Furthermore, the very limited opportunities for observation afforded by the weather obliged Halley to locate many of the stars in his map according to their distance from Brahe's fixed stars rather than by fundamental observation. In time, more accurate instrumentation allowed astronomers to recalculate with greater precision the position of both Brahe's stars and those related to them by Halley. One new constellation identified, which he named *Robur Carolinum* (Charles's Oak) in honour of King Charles II, is no longer acknowledged. The fundamental importance of Halley's work, however, remains undisputed.

sea-level and he reported to the Royal Society on a plant named Dianthum, which reproduced by novel means,[13] growing with

> a root on the Extremities of its leaves, and those sometimes will have others or Grand child plants (if such an expression may be allowed) growing out of their leaves; and that when the parent plant decays, the young ones fall to the ground and there take root, and so shift for themselves.

Nevil Maskelyne

In full knowledge of the limited accuracy achieved with his observation of the Transit of Mercury, Edmond Halley had been an early advocate for the importance of using the next Transit of Venus – anticipated on 6 June 1761, more than eighty years after the date of his visit to St Helena – to recalculate more precisely the distance of the Earth from the sun. His advice was widely heeded, leading to the Royal Society applying again to the Crown and to the EIC for support in making astronomical observations in the southern hemisphere. Their original intention was that two observers (Charles Mason and Jeremiah Dixon) should undertake one set of observations in Bencoolen,[14] while the Revd Nevil Maskelyne (1732–1811) and an assistant should do likewise in St Helena. For Maskelyne, a Cambridge mathematician and astronomer, the expedition to St Helena would be the first major opportunity of his scientific career. With the aid of a grant of £800 from the Lords of the Treasury, a suite of specially made instruments was commissioned with advice from James Bradley (who had succeeded Halley as Astronomer Royal), constructed with a degree of accuracy never before brought to bear on the island: two 2-foot reflecting telescopes, and a third for installation at Greenwich with which Bradley would make corresponding observations, two of the telescopes fitted with micrometers for the most accurate adjustments; a pendulum clock adjusted to compensate for the effects of temperature changes at St Helena; and an 18-inch quadrant with which the clock was to be regulated. Additionally, a 10-foot zenith sector was included to allow for a series of further observations on the annual parallax of Sirius.[15]

---

[13] Plants exhibiting this method of reproduction include the Maiden Hair Fern, *Adiantum caudatum*, one of the 'walking ferns'.
[14] In the end, serial difficulties resulted in Mason and Dixon getting no further than the Cape. Together they would later make their reputation by surveying the boundary that came to be known as the Mason-Dixon Line, separating Maryland from Pennsylvania.
[15] These details and much else here are taken from Harry Woolf's invaluable paper of 1956.

Maskelyne and his assistant Robert Waddington[16] arrived on 6 April 1761 on the East Indiaman *Prince Henry*. The St Helena government constructed an observatory for them on a ridge behind Alarm House (virtually all trace of which has now disappeared).[17] As Maskelyne later reported to the Royal Society, for the whole of the month preceding the appointed day, the view from the observatory was shrouded in cloud, so that he 'almost despaired of obtaining any sight of it at all'. In the event, he did 'obtain two fair views, though but of short continuance';[18] frustratingly, down in the James Valley the transit was distinctly seen by Waddington and by several other persons. With the brief window of opportunity afforded to him, Maskelyne – with limited success – 'measured the distance of the nearest limbs of Venus and the sun from each other, with the curious object-glass micrometer adapted to the reflecting telescope, according to Mr Dollond's ingenious invention',[19] but otherwise the moment had passed; when the clouds parted again there was no sign of Venus, although some miscalculation of the timing had suggested that the transit should still be visible. Maskelyne was compelled to wonder whether,

> on the occasion of the next transit, which is to happen eight years hence, it might not be convenient, that the observers should endeavour to place themselves on such parts of the globe, as that they may not see Venus on the sun's body, very near the horizon, but rather when they are both elevated to considerable heights.

His advice was doubtless of importance in sending Captain Cook and Joseph Banks to Tahiti to observe the 1769 transit. Maskelyne concluded his presentation to the Royal Society by remarking that 'if the late noble Dr Halley were now alive, he could not receive greater pleasure from seeing the observation of the transit of Venus undertaken by astronomers of different nations, conformably to his proposal'.

Maskelyne remained on St Helena until the following February, making the anticipated observations on the parallax of Sirius, a star that passes annually almost directly over the island. The disappointment of having been denied sight of the transit was compounded when inconsistencies emerged in his observations, later attributed to a defect in the astronomical sector with which he had been provided.[20] He also continued for a time to make observations on the island which later contributed to the first Nautical Almanac, while he and his fellow surveyors from the Cape came together on the island to check the

---

[16] For Waddington's contributions to the expedition and his subsequent career, see Bennett 2014.
[17] Tatham and Harwood 1974, 503.
[18] Maskelyne 1761.
[19] Dollond had earlier published a preliminary description of his 'contrivance for measuring small angles' in *Philosophical Transactions* 48 (1753–54), 178–81.
[20] Howse 2004.

rate of the clocks used in their observations. Towards the end of the exercise the discrepancies noted between the clocks' performance at the Cape and at St Helena contributed to the understanding of the difference in gravitational pull at different latitudes.[21]

Maskelyne used the opportunities on the voyage home (as he had on the outward journey) to make accurate observations on lunar distance which, when published in tabular form, would greatly ease the problem of finding longitude at sea.[22] This work was later continued with a voyage to Barbados under the auspices of the Board of Longitude and in 1765 Maskelyne was appointed Astronomer Royal.

THE LADDER HILL OBSERVATORY

While earlier observers had made do with rudimentary shelters put up to protect them and their instruments from the elements, a more formally constituted observatory was erected in the 1820s at the initiative of Governor Walker (Figure 17). It was conceived as part of a wider initiative on the part of Walker, who found himself, in the aftermath of the death of Napoleon and in a period when the island's security was no longer at risk from any quarter, in charge of a large garrison with a great deal of time on its hands. His solution was to channel the men's energies by establishing a Military Institute for the study of sciences, staffed by personnel from within the island and with a broad curriculum determined to some degree by the inclinations and talents of the instructors.

The founding of an observatory and time office was settled on as a useful initiative within this programme. The projected observatory was placed in the charge of Manuel Johnson, a lieutenant in the St Helena Artillery and aide-de-camp to the governor, whose natural aptitude for astronomy had singled him out for the post.[23] Johnson was seconded in 1825 to the Cape, 'to take the advice of His Majesty's Astronomer there on the subject'. His mentor at Cape Town, the Revd Fearon Fallows, 'immediately supplied a plan for a small Observatory [and] recommended suitable instruments'; thereafter Fallows remained a loyal supporter of the St Helena project.

---

[21] Mason 1761. Halley found that the length of the pendulum on his clock had to be shortened relative to its length in England in order to keep time in St Helena, an observation that would prove of importance in Newton's attempts to determine the true shape of the Earth (Thrower 1981, 19).

[22] Maskelyne's engagement with meteorology had led him to undertake observations at St Helena on the variations occurring there in tidal levels. A pole marked with a painted scale and fixed in the seabed near Jamestown provided the basis for a long series of measurements recorded through the day and much of the night, some of them undertaken by Waddington due to Maskelyne's duties at the observatory. These too were duly reported to the Royal Society (see Maskelyne 1762).

[23] See especially Warner 1982.

Fig. 17 The Ladder Hill observatory, erected in the 1820s under Governor Walker and operated under the direction of Lieutenant Manuel Johnson.

Observatory of St. Helena.

With the benefit of this advice – and with a series of tests of cannon fire to establish the most stable soil on which to build[24] – an appropriate site was chosen on top of Ladder Hill, 700 feet above sea-level on the rise adjacent to Jamestown.[25] The foundations were laid in September 1826 (to a royal salute from the St Helena Regiment followed by a '*feu-de-joie*') and the building was completed two years later.[26] There were two rooms for the meridian instruments and another intended for a zenith tube, as well as working spaces for the observers. After a second visit by Johnson to see the progress on

---

[24] These involved observing the effects of the blasts as transmitted to basins of mercury placed at different locations; ultimately the foundations were laid on a bed of lava 600 feet above sea-level (Warner 1982, 392–93).

[25] Amongst the other fruits of this initiative was the construction of the 900-foot ladder from sea-level to the top of the cliff at Ladder Hill; the ladder was flanked by rails laid on inclined ramps, up and down which agricultural and military stores could be hauled, with the aid of a capstan and ropes. Although highly useful, the rail tracks suffered from lack of maintenance and eventually fell from use, leaving only the staircase (now known as Jacob's Ladder).

[26] The foundation stone (the only part of the building surviving) is inscribed 'HAEC SPECULA ASTRONOMICA CONDITA FUIT MDCCCXXVI GUBERNANTE ALEXANDRO WALKER' (this astronomical observatory was founded in 1826 under the governorship of Alexander Walker).

a similar installation being built at the Cape, observations were begun at St Helena towards the end of 1829, Johnson being assisted by George Armstrong and by Sergeant Ramage of the St Helena Artillery. The instruments provided, made by Messrs Gilbert of Leadenhall Street, included a transit instrument with a focal length of 62 inches and a 4-foot mural circle. The observatory also had a barometer, thermometers, and a highly accurate clock by Barraud.[27] The latter not only served in the timing of astronomical observations but provided the basis for time-keeping throughout the island and for visiting shipping: William Webster was impressed by the practice of firing a rocket from the observatory at eight o'clock every evening, by which means ships could reset their chronometers and, over several days, could check the rate of gain or loss of their instruments. He wrote of the observatory that it 'bespeaks the liberality of the East India Company', and that under the control of Captain Johnson, 'in point of neatness and efficiency cannot be surpassed'.[28]

Johnson was indeed remarkably assiduous, making valuable observations on transits of the Moon and Moon-culminating stars, the position of Mars and the solar eclipse of 1832, in addition to a study of the tides at St Helena. His major work, the *Catalogue of 606 Principal Fixed Stars in the Southern Hemisphere*, published by the EIC in 1835 and subsequently awarded the Gold Medal of the Royal Astronomical Society,[29] amply justifies the confidence placed in him. The result of observations made 'from November 1829 to April 1833', it almost doubles the number of stars seen by Halley on his more transient visit to the island a century and a half earlier.

Sadly, by the time Johnson's volume appeared the observatory was already doomed. With the reversion of St Helena to the Crown in 1834, a massive cost-cutting exercise had been initiated, as a result of which, on 29 February 1836, the observatory was closed, due – in the words of officialdom – to its 'uselessness and immense annual cost of £300'; the Crown Commissioners sent to review the situation on the island declared themselves unable to learn that the establishment of the observatory had been attended with any important result to science: both statements are a telling indictment of the Nelsonian blindness that characterized the exercise.[30]

---

[27] These details are from Johnson's Introduction to his *Catalogue of 606 Principal Fixed Stars* (1835).

[28] Webster 1834, I, 352–53.

[29] Sir John Herschell, who was in line to receive the gold medal in that year, is said to have generously stood aside in favour of 'this young and private adventurer in that very arduous field', and instead received the award the following year (Warner 1982, 403).

[30] For his part Johnson, now in his thirties, left the army and, with remarkable application, went on to take a degree at Oxford where he was subsequently appointed director of the Radcliffe Observatory.

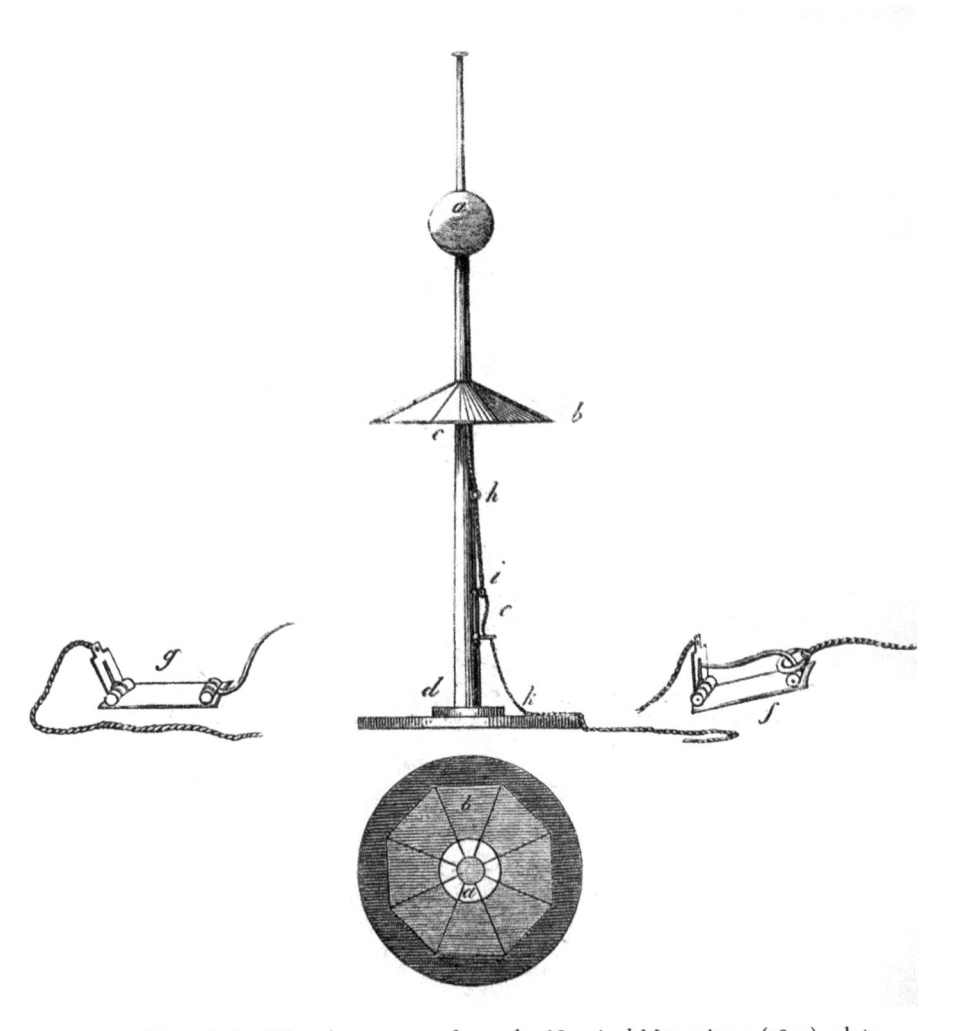

Fig. 18 Time-ball of the observatory, from the *Nautical Magazine* 4 (1835), plate opposite p. 658, which provided visiting vessels with the means of checking their chronometers.

Almost all the astronomical instruments were sent to Greenwich (and were subsequently dispersed), with only a few being retained to check the running of the clocks that did continuing service in the island's Time Office. One amenity to survive was a time-ball system similar to that instituted at Greenwich in 1833 – adopted in the same year at St Helena, making it the earliest in the South Atlantic – with the double purpose of allowing visiting vessels to adjust their chronometers and providing the resident population with a reliable time-check (Figure 18):

Scientists in transit: St Helena as a site for scientific investigation

The ball drops at mean noon, St Helena time, for the benefit of the inhabitants; and at one P.M. mean time at Greenwich, for the advantage of shipping. The ball is hoisted half-mast high five minutes before the time, and at one minute before to the mast-head.

A signal gun provided audible confirmation, and ships arriving at the island after one o'clock were able to request a repeat time-ball signal at the next hour by hoisting the Blue Peter to the main-topgallant mast-head.[31]

As for the observatory, the building itself was re-designated as a mess hall for the (much reduced) garrison, in which role it was seen in 1878 by Isobel Gill, wife of the astronomer David Gill,[32] who wrote of her visit to the observatory:

> I say Observatory – alas! it is so no longer. Fallen from its high estate, it is now the artillery mess-room, and in the recesses formed for the shutters of the openings through which Johnson's transit used to peep, they stow wineglasses and decanters, and under the dome they play billiards![33]

St Helena and the magnetic crusade

Just as the southerly position of St Helena recommended it to the astronomers, so the island's positioning at what was taken to be the area of least intensity of magnetic force on the globe singled it out for interest among those who concerned themselves with charting the variations in the Earth's magnetic field.[34] As with the precise positions of the stars, it was thought that (once they had been accurately charted) these variations, whose existence had long been known to navigators, might hold a key to locating the position of ships at sea. The constant changes taking place in the magnetic field were then little understood, but measurements undertaken at St Helena, when compared with those taken by early Portuguese navigators, showed that the magnetic field there had shifted by some 30 degrees in 250 years – an eye-opening discovery.

Edmond Halley was one of those who took an early interest in the phenomenon. In 1683 this led him to collect records of variations in the declination of the compass needle observed by mariners over the previous century. These in turn led to the first tentative conclusions that the variations in some way reflected the changing relationships between the 'shell' (or mantle) of the Earth and its core, although understanding of the mechanics involved remained rudimentary.

---

[31] *Nautical Magazine* 4 (1835), 658–60. The importance to mariners of the time-balls installed at several points throughout the Atlantic is discussed by Kinns 2021.
[32] Gill, later head of the Royal Observatory at Cape Town, used the site of the observatory as the base for his determination of the longitude of Ascension Island.
[33] Gill 1878, 26.
[34] Cawood 1979.

A further observation made by Halley has been taken to have contributed to Sir Isaac Newton's work on his theory of universal gravitation. At his arrival on the island, Halley had found it necessary to shorten the length of the pendulum on his clock in order for it to keep the same time as at Greenwich, a phenomenon for which he later sought an explanation from Newton. In July 1686 the latter responded that this necessity must have been caused by a lessening of the intensity of gravity at St Helena relative to London, supporting his deduction (later confirmed in the *Principia*) that the shape of the Earth must be that of an oblate spheroid rather than a perfect sphere.[35]

As master of the ship *Paramore*,[36] Halley again set sail from Deptford in October 1698 on what should have been a two-year voyage, intended 'to incompass the whole Globe', with several stated aims:

> ... the describeing and laying downe in their true Positions, Such Coasts, Ports and Islands, as the Weather will permitt ... and also to endeavour to gett full information of the Nature of the Variation of the Compasse over the whole Earth, as Likewise to experiment what may be expected from the Severall Methods proposed for discovering the Longitude at Sea.[37]

That expedition ended prematurely, but a second, twelve-month voyage, starting in September 1699 and focusing on the Atlantic as far south as the Antarctic ice fields, brought Halley briefly back to St Helena in 1700, when the *Paramore* spent three weeks at the island. The major achievement of the voyage was the chart compiled by Halley, which showed the variations in the magnetic field within the North and South Atlantic as isogonic lines. The chart was completed very quickly,[38] being presented to the Royal Society within two months of the vessel's return to England and published in engraved form in 1701 (Plate 18).

Henry Foster's observations with the pendulum

In his *Narrative of a Voyage to the Southern Atlantic Ocean ... in H.M. Sloop Chanticleer*, William Webster, the ship's surgeon and *de facto* naturalist, records the arrival of the vessel off Jamestown on Boxing Day 1828. The following morning the master of the *Chanticleer* and leader of the three-year expedition on which the

---

[35] Howarth 2007, 205. Halley also carried out a number of gravitational experiments on the island involving use of the pendulum. Gravitational pull at any one site was recorded by this method as a function of the periodicity of the pendulum compared to the length of the swinging arm (the time taken for the pendulum to swing being inversely proportional to the local acceleration due to gravity).

[36] *Paramore* was an Admiralty ship – perhaps the Royal Navy's first purpose-built research vessel – of the type known as a pink, quite small at 52 feet in length overall but commodious below decks, flat-bottomed to allow for inshore working, with a crew of twenty and mounting six guns.

[37] Commissioners of the Admiralty to the Navy Board, 12 July 1693, quoted in Cook 1998, 261.

[38] In fact the chart also incorporates some readings from Halley's first voyage.

company was engaged,[39] Captain Foster, went ashore 'for the purpose of finding a good position for making the pendulum experiments'.[40]

Henry Foster (1796–1831) was already an experienced participant in scientific surveys: he had gained admission to the Royal Society in 1820 for observations with the pendulum and served on expeditions to the Arctic on three occasions between 1823 and 1827. The experiments conducted there, on magnetism, refraction, and the velocity of sound, together with the astronomical observations and survey work he undertook, won him the society's Copley Medal and command of the British Naval Expedition that would take the *Chanticleer* on a circumnavigation of the southern hemisphere.[41]

Foster's pendulum observations undertaken on St Helena would contribute to a mosaic of similar geodetic data designed to give precision to the mapping of the world. In addition to those observations (aimed specifically at establishing with greater precision the ellipsoidal form of the Earth),[42] Webster mentions the accurate determination of longitude by means of chronometers as one of the expedition's tasks. The quest for accuracy in this matter explains why the thirteen-gun salute from the shore that greeted the *Chanticleer*'s arrival at Jamestown was returned on their behalf by another naval vessel then in the roads, the *Espoir*, 'in order that the rates of our chronometers might not be affected by the concussion'.[43]

Two observation points for swinging the pendulum were established by Captain Foster, the first at the 'castle or governor's house' in Jamestown, and the second in a guard-hut at the western extremity of the fort at Lemon Valley. While Foster undertook the exacting measurements at the latter, 'the young gentlemen of the Chanticleer were located on the hills in tents, employed in making magnetic observations'.[44] Multiple readings of periodicity were obtained using several carefully identified pendulums – two of brass (described as 'invariable') and two of copper ('convertible'); the expedition also carried an iron convertible pendulum, but no readings were recorded from it at St Helena.[45] No doubt, however, valuable data were added here to the expedition's observations on the powerful effects of

---

[39] Reassigned as a survey ship for the purposes of the voyage, the *Chanticleer* had already completed a trans-Atlantic voyage that took in the South Shetland Islands and South America before rounding Cape Horn for New Zealand and the Cape of Good Hope; from St Helena the ship proceeded to South America (where Captain Foster was drowned in Panama in a canoeing accident). Following her return to England the *Chanticleer* was scheduled to undertake survey work in South America but was found to be no longer seaworthy and was replaced by the *Beagle*.

[40] Webster 1834, I, 341, 380.

[41] Scott Polar Research Institute, Foster collection, https://archiveshub.jisc.ac.uk/search/archives/8c5b44fd-31e1-32b8-b518-61f5034d2725.

[42] Fogg 1992, 73.

[43] Webster 1834, I, 341.

[44] Ibid., I, 380; II, appendix.

[45] Some 20,000 readings were logged in the course of the voyage of the *Chanticleer* (Ibid., II, 212).

local attraction, 'which baffles all our efforts to deduce the true figure of the earth from pendulum experiments made at a few places only'; it was quite possibly here too that the realization emerged that the force of gravity 'seems to be greater in islands situate at a distance from the main land than on its continents', particularly islands that are 'mostly volcanic and consequently formed of dense materials'.[46]

THE OBSERVATORY AT LONGWOOD

In February 1840 building started on the site of a new magnetic and meteorological observatory at Longwood[47] – one of four such centres established by the British (the others being at Toronto, the Cape and Van Diemen's Land, now Tasmania) in response to Alexander von Humboldt's call for a world-wide chain of such centres.[48] The observatory was established (along with those at the Cape and at Hobart) at the initiative of the Royal Society by James Clark Ross, in the course of his Antarctic expedition of 1839–43; later it was administered by Major (later Lieutenant-Colonel) Edward Sabine (formerly secretary and afterwards president of the society) from his headquarters at Woolwich.

Responsibility for establishing the observatory had been deputed by Sabine to a young officer of the Royal Artillery, Lieutenant J. H. Lefroy, who arrived in 1842 on HMS *Terror* (one of the ships of Ross's expedition); Lefroy went on to have a distinguished career in the diplomatic service while retaining his interest in magnetism. It was he who settled on the site at Longwood and presumably he would have played a large part in the design of the observatory. The principal chamber where the magnetic apparatus was located was an octagonal room – lost when the building was converted to other purposes in 1849, although one or two features (including a slab of York stone that formerly supported the magnetometer, now re-used as a step) are still to be seen.[49] The former layout and the distribution of the equipment are preserved in a contemporary ground plan (Figure 19). The observations compiled by Lefroy and by his successors (named Smyth and Clark)

---

[46] Cf. Hinks 1944, 220: 'On those volcanic islands which rise steeply from deep ocean the force of gravity, as determined by the pendulum, is generally in excess, the excess being about that amount which can be attributed to the mass of the island standing above the ocean floor.'

[47] Longwood was chosen as a suitable site because of the considerable depth of soil there, which would insulate the instruments to some degree from the effects of the underlying rock on the magnetic readings.

[48] In a letter to the president of the society (dated 23 April 1836 and reproduced as an appendix by Malin and Barraclough 1991), Humboldt had suggested that, recording stations having been established 'from Paris to China', the study of magnetic variation could now be served if the society were to establish a number of stations in the temperate zone of the southern hemisphere; St Helena was one of the locations specifically singled out as being propitious for such an observatory. See also Cawood 1979, 509, 513, 516.

[49] Tatham and Harwood 1974.

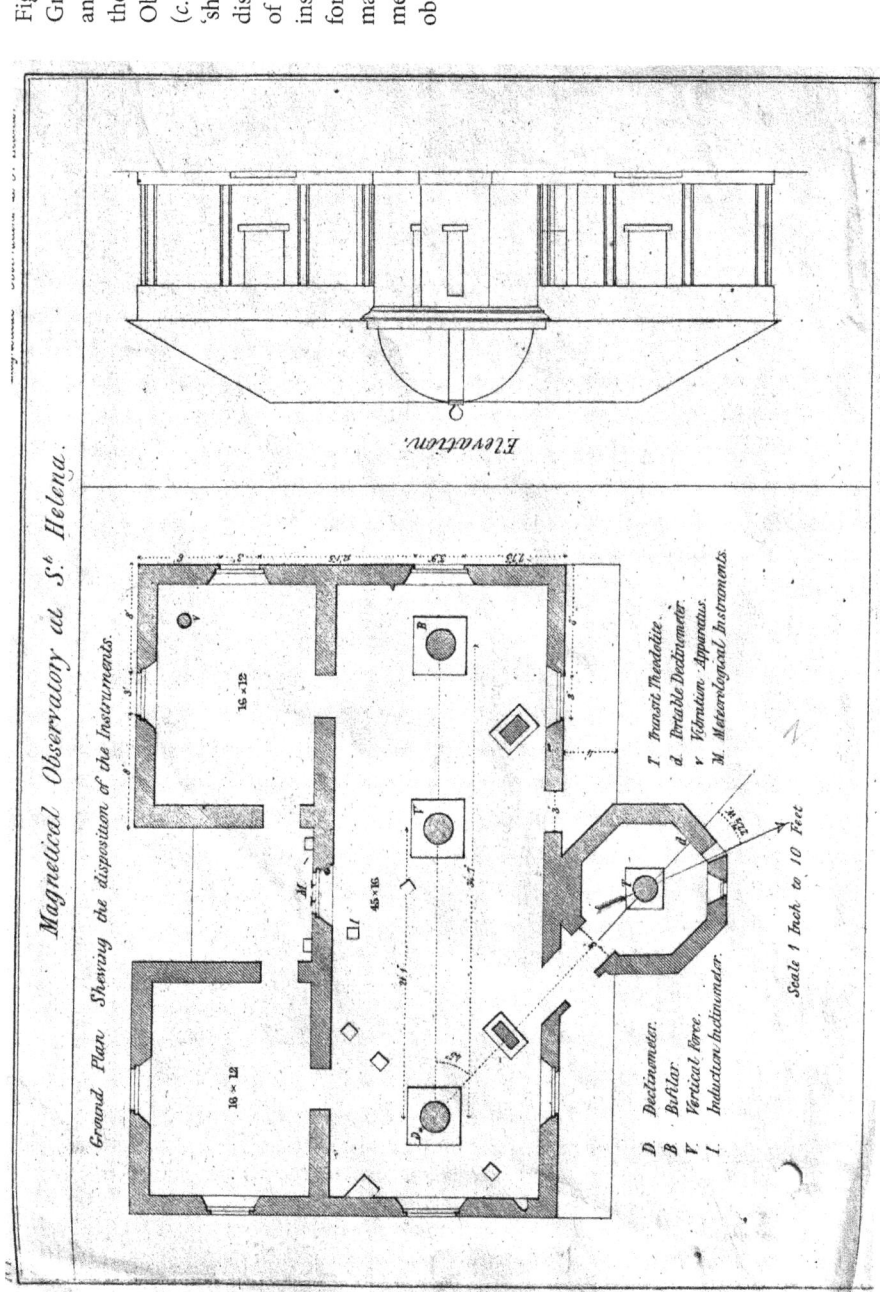

Fig. 19 Ground plan and elevation of the Longwood Observatory (c.1847), 'shewing the disposition of the instruments' for undertaking magnetic and meteorological observations.

were submitted to Sabine[50] and were incorporated by him in his *Observations made at the Magnetic and Meteorological Observatory at St Helena*, published in 1847, which includes descriptions of the instruments (as well as tabulating many of the observations made with their aid).

The observatory was again pressed into service when in 1890 the United States Scientific Expedition to West Africa (USSEWA), aboard the USS *Pensacola*, anchored at St Helena for nearly three weeks, during which the surveyors registered two further pendulum readings. One of these was carried out at Longwood, the other close to the Castle at Jamestown.[51]

Wind, weather and tide at St Helena

Much of Halley's later work on, for example, the mechanism of the trade winds and on magnetic variation, was founded on his experience on the island. Perhaps unsurprisingly, his confinement on a frustratingly cloudy and rainswept hilltop observatory, so damp as to render his notepaper impossible to write on, turned his mind to the mechanisms that generated all this moisture. Ten years after his visit to the island, by dint of controlled experimentation back in London, he developed his ideas to form what has come to be known as the hydrologic cycle, in which evaporation from the oceans and subsequent precipitation were shown to be in balance. The results of his investigations were communicated in a series of papers published in the *Philosophical Transactions* of the Royal Society.[52] The theory, he claimed, was not a 'bare Hypothesis' but was firmly founded on experience gained at St Helena.

More detailed observations of the weather (particularly rainfall) were carried out by members of the USSEWA during their sojourn on the island. The conclusion they reached was that the moisture was derived from cumuli formed by the trade wind impinging on the bulk of the island rather than from clouds formed by the relative heat of the ground surface. Rainfall on the island was judged as 'the best index to the average movement of the air, and as depending therefore, like the movement of the air, on the meteorological conditions over a large surface of the ocean'.

Air-flow was similarly thought to lie behind another of the phenomena frequently observed at the island, namely the occurrence of large 'rollers' on the leeward side that posed a considerable hazard to unprepared shipping. The

---

50   The reports survive in the archives of the Meteorological Office at Bracknell.
51   Todd 1890. Evidently the surveyors had contemplated further determinations at Foster's former station at Lemon Valley, but Todd tells us (pp. 8–9) that this plan was abandoned as impractical. The team also carried out measurements of gravity, temperature variation, barometric pressure and meteorology.
52   See Biswas 1970.

catastrophic force that could be unleashed by this unpredictable phenomenon was dramatically demonstrated on 18 February 1846 (see Plate 16):

> ... in the night the sea suddenly rose higher than had ever been known before and huge waves or 'rollers' broke upon the shore, so that at daylight next morning the sea opposite the town was one sheet of white foam and yet there was not a breath of wind. The Road was filled with shipping at the time, including eighteen slavers which were lying at anchor waiting to be broken up. At eleven o'clock the same morning, one of these, the Decobrador, was bodily lifted from her anchors and thrown broadside on top of another slaver, the Cordelia, and both were swept by the huge seas and deposited high and dry in front of the sea-guard gate. Altogether thirteen vessels were wrecked and smashed into pieces by the time the sea subsided the same evening.[53]

The surveyors from the *Pensacola* expedition had observed other (lesser) rollers and attributed them initially to the residual effects of distant winds, but on leaving St Helena and setting up an observation post on the top of Cross Hill, Ascension Island, they found their high vantage point provided them with a different hypothesis. According to that theory the topography of both small islands, each set in a wide stretch of ocean swept by the trade wind, is such that the swell caused by a wind blowing constantly at no more than Force 4 for a couple of days is deflected to the leeward side in the form of a roller – a double roller if deflected both clockwise and anticlockwise. 'It will at once be a matter of surprise that rollers are peculiar in their severity at Ascension and Saint Helena', the observers noted, 'but ... the severity of the rollers depends first on the shape of the island; second, on its size; third, on the location and character of the shoals which surround it, all taken in connection with the length and height of the original swell'. They could think of only one other island (St Paul de Noronha, in the mid-Atlantic off Brazil) where the rollers are equally conspicuous.[54]

Anecdotal evidence – or rather the accumulated observations of generations of seafarers – had led the Admiralty's *Africa Pilot* to observe by the 1960s that the rollers, 'which break with great violence on [the island's] lee sides, and arrive without any apparent warning', were in fact 'progressive undulations caused by distant storms, in the Atlantic oceans of either hemisphere'. The source is quoted by D. E. Cartwright, who undertook a detailed tidal survey in the late 1960s; one of the main objectives of his expedition had been to analyse the origin of the rollers, and although none occurred during the period of his survey he concluded that it was 'extremely likely that genuine "rollers" merely consist of large amplitude swell, the sudden shoaling close to the island producing the sudden increase in height which gives the impression to local

---

[53] Gosse 1990, 316–17.
[54] Todd 1891, 565.

mariners of a disturbance "arising from the depths of the ocean".[55] More recent research also favours such an origin: following the installation of sub-surface pressure-sensors at St Helena and Ascension Island in 1999, a tide monitoring exercise recorded an unusually large deep-ocean swell that the researchers were able to attribute by means of wave modelling to the effects of a North Atlantic hurricane that had broken the previous week.[56] Their origin would now seem to be settled, but their arrival remains as unpredictable and potentially as devastating as ever.

A succession of naturalists

An early recognition of the special status of St Helena's vegetation, the result of its natural remoteness and of the all-too-visible effects of human intervention, rendered the island of particular interest to early naturalists. Some of those who made significant contributions to understanding of the island's natural history are discussed here: there must have been many more transient visitors whose contributions are less easy to detect. Mention might be made of Philippe Welle, *Hofgärtner* on the staff at Schloss Hetzendorf, a satellite establishment of the imperial palace of Schönbrunn, who arrived in 1816 in the company of Baron von Stürmer, the commissioner appointed to represent Austrian interests during the detention of Napoleon. Welle spent six months on the island and did some collecting with a view to stocking the gardens at his master's lodgings at Rosemary Hall, but seems to have left little trace of his presence beyond attracting the ire of the governor, Sir Hudson Lowe, by communicating a message from the mother of one of Napoleon's servants to her son. The incident brought about Welle's premature expulsion from the island, accompanied by unspecified plants destined for the princely gardens of Austria.

JOSEPH BANKS

Joseph Banks (1743–1820) was already a significant figure in the British scientific community when he spent four days on the island in 1771, on the return voyage of Captain James Cook's *Endeavour* from its triumphant circumnavigation via Cape Horn, the Pacific Islands (where he had observed the Transit of Venus on Tahiti), New Zealand, Eastern Australia and the Cape of Good Hope. Banks's experience as naturalist on the voyage (accompanied by Daniel Solander) would cement his status in England and contribute to his election as president of the Royal Society and to the award of a baronetcy in 1781.

The prospect of visiting the island had evidently proved intriguing for Banks the naturalist:

---

[55] Cartwright 1971, 628–32.
[56] Vassie, Woodworth and Holt 2004.

Scientists in transit: St Helena as a site for scientific investigation

> Secluded as this rock is from the rest of the World by seas of immence extent it is difficult to imagine how any thing not created in that spot could by any means arrive at it; for my part I feel more wonder in the finding of a little Snail on the top of the Ridges of St Helena than in finding people upon America or any other part of the Globe.

But Banks was also a landowner with considerable estates in Lincolnshire, and the critical eye he brought to St Helena was that of a rural businessman as well as a scientist. The island disappointed him in both respects. Certainly he judged that the island's paradisiac qualities, extolled over the previous two centuries, had been ill-served by the settlers. In terms of natural resources St Helena was, he considered, better endowed than the Cape but had been brought to a degenerate state through endemic mismanagement. He was unsparing in his comparison of the two populations:

> In short the Cape of Good Hope, which tho by nature a mere desert supplys abundantly refreshments of all kinds to ships of all nations who touch there, contrasted with this Island, which tho highly favoured by nature, shews not unaptly the Genius's of the two nations in making Colonies: nor do I think I go too far in asserting that was the Cape now in the Hands of the English it would be a desert, as St Helena in the hands of the Dutch would as infallibly become a paradise.

Of the residents, who depended for much of their living on supplying visiting ships with refreshments, he wrote that, 'to their Shame be it spoken they appear to have by no means a supply equal to the extent as well as the fertility of their soil, as well as the fortunate situation of their Island seem to promise'. Garden vegetables and fruit grew well here, he noted, but were 'so far from being in plenty so as to supply the ships who touch here a scanty allowance only of them are to be got ... and tho there is a market house in the town yet nothing is sold publickly'. The islanders' pastures lay 'as much neglected as their Gardens', and in general the practice of visiting captains in paying extravagant prices for what they did receive he considered had 'inspird the People with a degree of Lazyness'.

Banks was told that some attempt had been made about forty years previously to grow barley, and although the results had been promising, 'its cultivation was however suddenly drop'd, for what reason I could not find out, and since that time has never been attempted'. To repeat a passage quoted earlier, even yams, which 'they chiefly depend upon to supply their numerous slaves with provision', were not 'cultivated in half the perfect[i]on that I have seen in the South Sea Islands, nor have they like the Indians several sorts many of which are very palatable, but are confind to only one and that one of the Worst'.

If cultivated plants were thin on the ground, significant populations of wild plants proved even more scarce: the island's famous cabbage trees were now restricted to the higher ridges, although gumwood continued to grow at lower altitudes; the ebony tree had been brought almost to extinction, while 'other species of trees and plants which seem to have been originally natives of the island are few in number'.[57]

Whatever these shortcomings, the strategic position of the island impressed itself on Banks's mind and in later life he promoted its role as an important node in the network he envisaged for the international exchange of useful plants, forming a way-station in their transfer from the Indo-Pacific region to Europe and to the West Indies.[58] Perhaps the most dramatic manifestation of this exercise in 'green imperialism' was the arrival at the island in 1792 of Captain William Bligh with a shipload of breadfruit plants, *en route* from Tahiti to St Vincent (see below).

Richard Grove convincingly identified Banks's experience of the island at this time as the origin of a view, articulated later in life, of St Helena as a vision of paradise lost, contrasting both with 'the sensuous and unspoilt image of Otaheite and with the artificial paradise of the Cape'.[59] In particular, the beneficial effects of the flow of ideas from naturalists such as Clusius (Charles de l'Écluse) and Henrik Adriaan van Reede to the Dutch colonies overseas are singled out for approval by Grove, and contrasted with the comparatively unenlightened regime under which St Helena was administered. In time Banks would come to adopt something of the role of mentor towards the island's would-be improvers: his view of the regrettable depletion of the ecosystem was always balanced by his eye for opportunities for expansion and diversification of agriculture, fruit and vegetables. In particular, the correspondence he struck up from 1784 with the newly appointed Huguenot governor, David Corneille, proved at first more fruitful to him than his correspondence with the Court of Directors. In time, however, Banks came to be consulted by the EIC on a wide range of matters that reflected the range of his own encyclopaedic interests, and he also used his connections with the Royal Gardens at Kew to the benefit of the island. The founding of the botanic garden on St Helena in 1787 may be attributed directly to his influence.

JOHANN REINHOLD FORSTER

Banks had served as naturalist (or rather, had personally financed the entire investigation of natural history) on Captain Cook's first voyage to the South Seas. Cook called again at St Helena four years later, on the homeward leg of a three-year circumnavigation in the *Resolution*, on which the naturalist's role was fulfilled by Johann Reinhold Forster (1729–98). The *Resolution* found itself becalmed just short of the island on 16 May 1775 and had to lower boats to tow her into James Bay, finally dropping anchor at midnight.

In some ways Forster's observations on the island were more acute than those of Banks, and certainly he seems to display more understanding of the difficulties

---

[57] Banks 1962, II, 269.
[58] Ly-Tio-Fane 1996.
[59] Grove 1995, 342.

under which the population laboured. He admired the fields 'regularly divided by stone-enclosures', the agreeable houses of the population, 'surrounded by gardens, orchards & plantations', and the 'genteel' lodging of the governor. On the matter of the paucity of green vegetables grown, he notes that 'Cabbages grow here well in other Seasons but when they are most wanted at the return of the Shipping from India, they are devoured by small Caterpillars & other Insects'. He found that potatoes, peas, beans, carrots, lettuces, artichokes, asparagus and cucumbers all grew well, but suffered a disadvantage:

> The European Garden Seeds must be renewed, because they degenerate, for which reason the Company has wisely ordered that each Ship touching at the Cape should import from thence to St Helena the value of 5£ Sterl. in Garden Seeds, which is commonly complied with, but the choice of the Seeds is left to the people at the Cape; who send very little Pease, Beans & Kidney Beans, or of other useful things & make the rest up of Mustard, Cresses & Raddishes, things which now are growing wild on the Isle.

There were other difficulties: 'Indian-corn', he noted, 'formerly has been sowed in greater Quantities, but is now almost lost in the Isle: but it is said the numerous Rats & mice destroy vast Quantities of it'.

Livestock throve on the 'fine Turf & continual Grass' in the pastures, but they too faced a problem:

> ... we suspected the Isle could feed at least as many Cattle again, but we were told by the Inhabitants that the Grass must be spared now towards their Winter Season, For Grass would not grow again up during Winter, & if they had not spare Fields with Grass to put the Cattle in, they starve in Winter.

Nonetheless, Forster felt that the livestock population could usefully be increased from its current 2,600 head to 3,000, 'without distressing the Cattle', and that lucerne and sainfoin could be beneficially introduced to the pasture.[60] He makes no mention of horses, but advocates the importation of asses from Senegal, which seem 'best calculated to this Climate, both for burdens and for riding'.[61]

Amongst the game-birds on the island Forster notes the presence of partridges and European pheasant, the latter recently introduced by Governor Skottowe along with

> the Guinea-Hen & will in a few Years grow very numerous, for at present there is a Penalty of 5£ Sterl. for killing them, till they multiply to such a degree that

---

[60] Johann Reinhold's son, George Forster, who accompanied him on the voyage, was of the opinion that clover and snail trefoil could be planted to provide more substantial food for the livestock (Forster 2000, II, 665).

[61] Gosse (1990, 132) equates these 'asses' with donkeys, but perhaps the term refers to the 'wild asses from Senegal' favoured by the Forsters (the ancestors of European donkeys) or to mules.

they may become common Game, which period in all probability is not very far off, for we saw a good many in this Excursion.

In his comments on the wild flora Forster notes the usual eye-catching native varieties, notably gumwoods and cabbage trees, but seems sanguine about the introduction of foreign species. He notes, for example, that at the governor's house, 'A large American willow beard live Oak makes a fine Appearance, & is a proof that its growth would succeed in this Isle.'

JAMES RENNELL

James Rennell (1742–1830) had already completed an exceptionally successful career at sea and latterly in India (where his service with the EIC had culminated in ten years as Surveyor-General) when he arrived at St Helena aboard the *Earl of Ashburnham* in 1777, homeward bound from Calcutta at the completion of his service. Characteristically, he had made use of the journey by attempting to chart the oceanographic currents encountered on the voyage: his *Chart of the Bank of Lagullus*, published in 1778, was the first successful attempt at mapping what is now known as the Agulhas Current, whose impact on the biological diversification of St Helena has been mentioned above. In a long life back in England, Rennell continued to build on the theories on ocean currents he had begun to formulate on the voyage. By correlating reams of data extracted from ships' logs, he was able to give form to his emerging theories, resulting in the appearance of his *Investigation of the currents of the Atlantic Ocean and of those which prevail between the Indian Ocean and the Atlantic*, published posthumously in 1832, a volume regarded as forming the historical basis of ocean circulation studies. He was also a founding spirit of the Royal Geographical Society.

From a cartographic viewpoint, Rennell's brief stay on St Helena provided no significant initiatives, but it did see the birth of his daughter Jane, who in later life (as Lady Rodd, wife of Admiral Sir John Tremayne Rodd) would publish in collaboration with John Purdy her father's hydrographic charts and produce new editions of his principal works.

CAPTAIN WILLIAM BLIGH

The year 1792 saw the arrival at St Helena of an extraordinary cargo, carried aboard the Royal Navy sloop HMS *Providence* under the command of Captain William Bligh (1754–1817); the *Providence* was accompanied by a smaller escort vessel, HMS *Assistant*. In truth, the island was no more than a staging post on a voyage whose primary mission was to carry living specimens of breadfruit plants gathered in Tahiti, for transplantation to the West Indies. The venture was the brainchild of Sir Joseph Banks and its aim was to provide a reliable source of nutrition for the enslaved populations of Britain's West Indian possessions, where supplies had been rendered scarce by a succession of hurricanes and by an interruption of trade due to the American War of Independence.[62]

The choice of Bligh as commander of the expedition, within months of his exoneration by court martial for his loss of the *Bounty*, proved a wise one. A talented navigator and surveyor, he was familiar with the seas both in the Pacific and in the West Indies. In order to facilitate the mission, the ship's company included two botanist-gardeners – James Wiles and Christopher Smith – who had been carefully instructed on the question of which plants to carry out from England for planting overseas in the course of the voyage – another of Banks's initiatives for improving biological and economic diversity – and on the means of collecting and conveying breadfruit and other plants for the West Indies. They were also to bring back plants for the Royal Gardens at Kew and for Banks himself, and were further permitted to collect specimens on their own account, provided it did not interfere with the expedition's primary purpose.

The two ships (the *Assistant* under the command of Lieutenant Portlock) had left England on 3 August 1791, calling at the Cape and at Van Diemen's Land *en route* to Tahiti. At each stop, plants from the cargo were dispensed and new species taken on board. Arriving finally at Tahiti in April 1792, the crew and the gardeners had set to gathering breadfruit in particular, so that by 26 May Bligh could record that: 'all the plants are now in charming order, and spreading their leaves delightfully. I have completed nice airy spaces for them on the quarter deck and galleries and shall sail with every inch of space filled up.'

After three months on Tahiti, the two ships had sailed with 2,126 breadfruit plants, 472 other plants and 36 'curiosity' plants on board. They put in at Timor, where further species were collected and surveys undertaken, before setting off to round the Cape in favourable spring weather and to reach St Helena on 17 December, with some 830 plants surviving the voyage – a number the gardeners hoped would 'exceed [Banks's] most sanguine expectation'.[63]

The expedition was welcomed by Governor Brooke, who received in return ten breadfruit plants in order to secure for the island 'a lasting supply of this valuable fruit which our most gracious king had ordered to be planted there'.[64] As a first duty, the company saw the new introductions safely planted, as recorded by Bligh:

> the principal plants were taken to a valley near [the governor's] residence called Plantation House and the rest to James's Valley. On the 23rd I saw the whole

---

[62] The account of events given here is indebted to the work of Dulcie Powell (1977).
[63] Powell 1977, 397.
[64] David 1993, 823. Having been successfully nurtured for a number of years, some (at least) of these same plants succumbed to drought a few years later, for the Annual Register for 1798 records: 'A scarcity of rain for three successive years had caused great mischief and want at St Helena. Several of the breadfruit plants which had been left by Captain Bligh on his return from the South Seas had fallen to decay' (quoted in Ashmole and Ashmole 2000, 62).

landed and planted; one plant was given to Major Robson, Lt.-Governor, and one to Mr Raughan, the first in Council. I also left a quantity of mountain rice seed here. The Peeah was the only plant that required a particular description. I therefore took our Otaheitian friends to the Governor's House where they made a pudding of the prepared part of its root, some of which I brought from Otaheite.[65]

The gardeners, Wiles and Smith, sent a more detailed account to Banks on Christmas Eve, describing (with justifiable pride) the condition of the plants they had nurtured so assiduously and which were destined specifically for St Helena:

> During the short time we have been here the Collection of Breadfruit and other plants has received much benefit from the fine Air which constantly blowes off the Land – Mr Portius[66] the Botanist had made every necessary preparation to receive the Portion of Plants you have been pleased to assign to them, and we thought it our duty to give you an account of the Plants left and collected at this Place; we have left 11 fine healthy Breadfruit Plants, one of them very large besides 12 others in a sickly State, some of which will most probably recover, altho' they would have all certainly died before we arrived at St Vince[n]t – 5 Ayyahs, 4 Rattahs, 2 Avees, 1 Mattee, 4 Ettows and 2 Peeah. Of the Timor Plants are left 4 seedling Nanches or Jacks, 2 Mangos, 2 Jambo mare, 4 Jambo armarvah, 2 Long Pepper, 2 Black Pepper, 2 Penang, or Beetlenut, 1 Lemon China, 1 Bughna Kanana a perfume, making in the whole 61 Plants; besides a Portion of two Species of Mountain Rice, and seeds of the Avee, Rattah, Candle Tree or Tootoo, Ettow, Mango and Peeah – We have collected 3 large Plants of the Fern Tree and fill'd 30 Vessels with curious and useful Plants for the West India Islands and His Majesties Botanic Garden at Kew – We are much obliged to Mr Portius for exerting himself in a particular manner to supply us with every different Plant on the Island worth notice – The Breadfruit left here were transplanted three days ago in a fine loamy Soil and excelent Situation, they came out of the Pots with the whole ball and appear not to have miss'd the moving.[67]

---

[65] 'Our Otaheitian friends' were Mydiddee and Pappo, who had sailed with Bligh from Tahiti and were warmly received during their stay on St Helena. 'An Account of the Root called Peeah & of its Use' (including the method of making it into a pudding) is included among the correspondence from Bligh to Banks, held at the State Library of New South Wales (https://www.sl.nsw.gov.au/banks/section-09/series-50/50-38-an-account-of-the-root-called-peeah-).

[66] Henry Porteous, superintendent of the garden, also kept a boarding house next to the garden. Writing of a visit to the garden, Lord Valentia (an amateur naturalist himself) writes that, 'although there is a botanist appointed by the India Company, has no pretensions to that title, as there has not been an attempt to collect even the indigenous plants of the island' (Annesley 1809, I, 14).

[67] Reproduced from Powell 1977, 398–99.

Later they would draw up a list of forty-eight plants (sixteen species) from St Helena that survived the voyage to be planted in St Vincent.[68]

The voyage remains perhaps the single most iconic illustration of Banks's plan for a world-wide exchange network in which famine in one area might be ameliorated by the introduction of new species from elsewhere, a process relying not only on expert seamanship and husbandry of the plants while on board but on the establishment of botanical gardens where the new introductions could be nursed and acclimatized. For St Helena the visit of the *Providence* provided a brief window into the world of 'green imperialism',[69] but for all its renown, Captain Bligh's introduction of the breadfruit to the West Indies seems to have enjoyed only limited success, for the slave population would not eat it (although curiously it is said to remain popular in Puerto Rico to this day).

WILLIAM BURCHELL AND 'THE NATURAL PRODUCTIONS OF ST HELENA'

Unlike the other scientists discussed here, the arrival on St Helena of the twenty-three-year-old William Burchell (Figure 20a) on 13 December 1805 was unsanctioned by any official body: indeed Burchell did not even have the EIC's permission to land on the island, but a bout of diplomatic illness brought him ashore and a kindly reception from the governor, Colonel Patton, secured him almost immediately an initial appointment as schoolmaster in Jamestown.[70] Patton evidently saw in the new arrival (son of the prosperous owner of the Fulham Nursery and Botanic Gardens, a former employee at the Royal Gardens at Kew and a Fellow of the Linnean Society) an ally in his ambitions to gain a systematic understanding of the natural history of the island and of the possibilities for its exploitation.

No greater enthusiast could have been found for every aspect of St Helena's natural riches. Burchell responded to the island's topography and vegetation with a scientist's eye and a romantic's heart: his journal, surviving in the University Museum of Natural History at Oxford (OUMNH),[71] records numerous expeditions on foot and on horseback to the interior, where he revelled in the richness of it all, opportunities that seemed destined to increase when, under a different governor, Colonel Beatson, he was appointed 'the Company's Naturalist'.[72] For example, on

---

[68] Listed in detail in ibid., 400.
[69] See Grove 1995 for the definitive exposition of this doctrine.
[70] Cleverly (1989, 6) mentions that Burchell's letter of appointment concludes with a reprimand for having landed illegally, but that seems to have been the end of the matter. See also McKay 1934, 481–82.
[71] A transcription of the full text, held in the Oxford University Museum of Natural History, is reproduced in Castell 2011. I am grateful to Dr Malgosia Nowak-Kemp for access to information on the holdings of the OUMNH.
[72] A salary of £200 was initially proposed for his full-time work, later increased to £250 and subsequently £300 (Castell 2011, 118–19, 147).

Fig. 20 A succession of naturalists: (a, left) William Burchell, (1781–1863); (b, below) William Roxburgh (1751–1815); Roxburgh visited St Helena on his way to retirement in Britain after a long service in India; the others were all on their way to making distinguished careers for themselves.

Fig. 20 (c, above) Charles Darwin (1809–1882); (d, right) Joseph Dalton Hooker (1817–1911).

11 October 1807 he describes an outing to Diana's Peak, the highest point on the island:

> As you advance within the paradise of verdure that surround the Peak it is impossible for any contemplative person not to feel the greatest delight that scenes of Nature, clothed in their wildest robe, and pure air, can inspire. You wind through the most romantic glens, sometimes covered with dark thickets, composed of the purple and white Cabbage tree, the wood, the thick leaved Cabbage tree and very tall trees of Furze; all of which are generally impenetrably connected by bushes of the Pumated-leaved Bramble. Or if any of these trees form little groves the ground beneath them is always strictly covered with a great variety of large ferns. In one of these woody glens I was struck with the beauty of a noble large Redwood tree, which rose exultingly above the thicket seemingly vain of its handsome large white flowers … As I got nearer to the top, my attention was more forcibly arrested by the most singular fern the Dickinsonia. I gazed with delightful surprise on this remarkable production. I scarcely dared to believe that it could be really a fern. Its robust stem, or rather trunk, and its height which sometimes equalled 20 feet, forbade my considering it as belonging to a tribe, which, instead of overtopping the surrounding trees, is rather contented to rise a little above the grass …

Repeated expeditions of this kind produced a rich haul of plants, whether for drying as herbarium specimens, for recording in systematic lists and sometimes for drawing, or for transplanting. Only the day previous to the outing cited above he had discovered on Sandy Ridge 'many plants I had never seen before, particularly a species of Lobelia a Campanula, the Purple Cabbage tree, a noble specimen of Asplenium, a Lonchitis, and a Marchantia', and so on. The end result was a list of over 200 wild plants recorded in the very first systematic flora produced for the island.[73]

In addition to his detailed inventory of the flora, two aspects of the observations made by Burchell are of exceptional importance. Not only was he the first to distinguish a number of plants as endemic to the island, but his growing familiarity with every part of the terrain allowed him to recognize that many species were in decline or had already suffered extinction as a result of human impact on the ecosystem. Following an expedition to Little Stone Top on 8 December 1807 (Plate 19) he wrote:

> I have noticed that all the way we have come is strewed with the decayed remains of trees and shrubs which formerly have nearly covered all these hills, and is a melancholy proof that the growth of wood and verdure at St Helena is decreasing, nor is this to be wondered at, when we hear that, without providing for posterity

---

[73] In the Archives at the Royal Botanic Gardens, Kew, are two of Burchell's volumes of relevance here: his 'Flora Insulae Sanctae Helenae' in tabular form and a 'Flora Heleniana', containing manuscript descriptions of some of the plants in the tabular list.

by young plantations, the soldiers and inhabitants have been suffered barbarian like, to cut down the trees with a wanton waste, only making use of the stems and thick branches, leaving the brushwood behind. Almost everyone can tell me that they can remember thick and almost impassable groves of trees growing on those hills, which now offer to the eye nothing but a cindery barrenness.[74]

Shortly afterwards, when his companion accidentally shattered a mature native tree by dislodging a boulder on a steep hillside, he evidently felt personally implicated:

> I viewed the demolition of one of these ancient Gum Trees (shrubs) with a superstitious concern, and the feeling of a fellow creature, for in all probability, unless St Helena should be deserted, these trees should never again be suffered to attain so great an age, and (as this tree is peculiar to the Island) this was sacrilegiously destroying the largest of the kind that would ever again be in this world.[75]

He was moved too by the many ruinous farmsteads he encountered, the former occupants of which had evidently abandoned cultivation in favour of a more comfortable life in trade in Jamestown.

An integral role in Burchell's plant-hunting activities came to be played by the botanic garden in Jamestown, which he established – or more strictly re-established, since there already existed a garden of rudimentary character on the site – adjacent to the Castle and overlooked by other buildings of the East India Company. During the 1790s a certain amount of landscaping and enclosure of what had hitherto been wasteland had been achieved by use of soldiers on punishment detail in lieu of flogging, though the combination of poor soil and low rainfall rendered its success perpetually precarious.

Burchell's suggestion that a more formal and purposeful botanic garden should now be established (Figure 21) had received immediate support from Governor Patton. The initial stock came from within the island, gathered either from the gardens of the more enterprising inhabitants – notably at Plantation House, the governor's country residence – or from the wild. A batch of plants arrived from Plantation House on 31 March 1807, 'the first received at the Botanic Garden', and when a second donation arrived the following day Burchell already found himself under stress: 'as the carpenter had not time to make my seed boxes I sent to him for some boards and nails, being determined to make them myself, rather than wait any longer'. Later entries in his journal record the arrival of other garden plants including gladioli and ixias – corn lilies native to southern Africa, evidently introduced to St Helena at some earlier date.

---

[74] Castell 2011, 72. Later, on 2 April 1810, Burchell was incensed at a decision of the government to raise the price of coal, a move he feared would 'prove of irreparable detriment to the Island, as it will cause a greater consumption of the native fuel. The consequence of which must infallibly be the cutting down of every tree and bush that now give shelter and shade the pasture.'

[75] Castell 2011, 73-74.

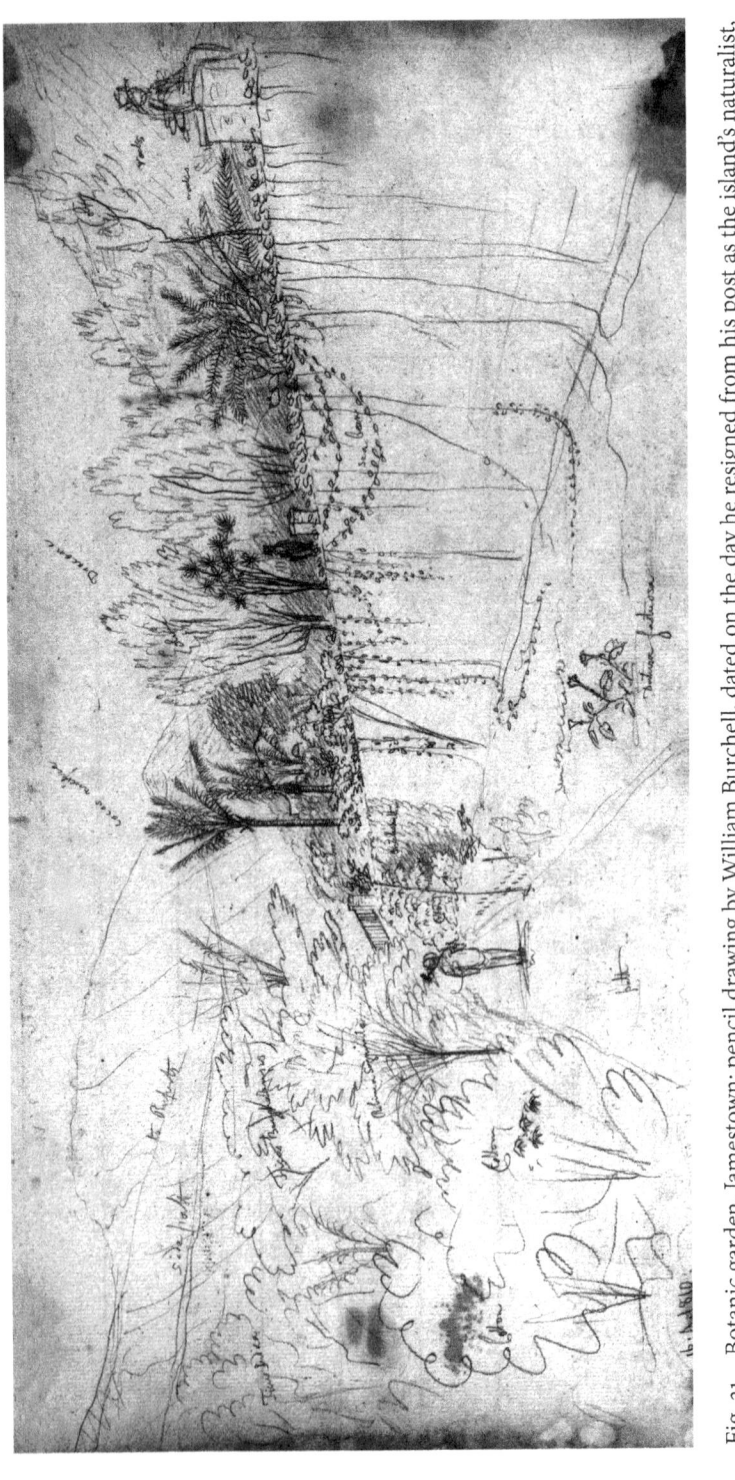

Fig. 21 Botanic garden, Jamestown; pencil drawing by William Burchell, dated on the day he resigned from his post as the island's naturalist, 16 April 1810.

Scientists in transit: St Helena as a site for scientific investigation

Overseas plants took on an increasing importance when the garden began to act as one of the more regular centres of acclimatization for plants making their perilous way to Britain (notably to the Royal Gardens at Kew) from India, the Far East and the Pacific.[76] A whole new world was opened up for Burchell by these contacts, first witnessed by the arrival on the island on 17 April 1807 of a 'Mr Drummond of Canton',[77] who had 'brought some plants with him intended for the Kew Garden', and to whom the governor 'had mentioned our Botanic establishment here and proposed for some to be left at St Helena'. Drummond, who 'seemed very desirous of promoting [the garden] by every means in his power', immediately offered some seeds and said he would write 'to Mr Roberts, his successor at Canton, and Mr Kerr the gardener at Macau, and to Mr Duncan at Pinang'. On 23 April Burchell was 'very busily employed all day in getting the Kew plants on shore' from the East Indiamen *David Scott* and *Ceres*. Three days later, and evidently impressed by what he had seen, Drummond told Burchell that he was considering 'leaving all the plants here in order to establish them on the Island whence they could at any time be sent to England', since the destination of his own fleet remained uncertain at the time. Within a week the contents of the *David Scott*'s plant cabin had been landed and an exchange had been instituted of 'the healthiest [plants] I had on shore for the sickly ones' in the *Walmer Castle*; the seeds destined for Sir Joseph Banks at Kew were also landed so that Burchell could take a part of them.

Other exotics followed: in March 1808 'Mr Rose (A Dutchman and late Member of the Council of Batavia) called on me with a basket of succulent plants sent to me from the Cape by Mr Hesse', sixty-seven of which were later planted in the garden and seeds from the same source were sown. On 13 June 1808 a letter arrived from William Roxburgh, superintendent of the botanic garden at Calcutta, together with 'a parcel of 130 sorts of seeds'. On 14 July 'a box of plants sent to me from China' arrived via the *Alfred*, 'but they were I fear all dead', followed on 11 August by, via the *Europe*, 'the bag of Larch cones sent me by Mr Drummond'. On 6 May 1809 he records that 'Mr Johnson called on me bringing some Ceylon seeds', for which Johnson received in return a parcel of thirty kinds of seeds and other specimens.[78]

---

[76] Three such gardens were established on St Helena – that in Jamestown reserved for coastal plants, at Plantation House for middle zone plants, and the third on Diana's Peak for those from higher altitudes (see McCracken 2022, 81). The scale of Burchell's ambitions for these botanical interventions can be glimpsed in his declared ambition that St Helena might yet 'become profitable instead of being an expensive possession': quoted by Cleverly 1989, 8.

[77] This must surely be Captain Charles Drummond mentioned in a letter of 29 December 1804 from Robert Kerr to Sir Joseph Banks: by Drummond's ship *Glatton*, Kerr is 'sending to India House seeds and drawings by a Chinese artist'; Kerr mentions that he has also heard from 'Lance' at St Helena, 'asking for Fruit-trees for that island', which had been sent.

[78] Burchell's sights were set on other sources apart from the Orient: on 24 March 1808 he commissioned a colleague sailing on the brig *Three Brothers*, bound for Rio de Janeiro, to have collected for him 'every possible kind of seed from the Brazils'.

Already, however, all was not well in the botanic garden. The scheme that had flourished under the sympathetic regime of Governor Patton,[79] had been thrown into confusion with the appointment as governor of Colonel William Lane – as acerbic, overbearing and narrow-minded as Patton had been enlightened. On 1 August 1807, Dawson, Burchell's gardener, brought news that Lane

> had been to the Botanic Garden where he had signified his intention of entirely altering all my plans and of overthrowing all my arrangements He said that let it cost him what it would, he would have some good pines [pineapples] out of that garden yet. I had just been making a trellace for the different twining exotic plants that I had collected and reared. This he said would just do for him to plant vines against and it appears that he is resolved to undo whatever the good and excellent Governor Patton strove to accomplish.

Although Burchell did what he could to oppose Lane's plans, he could not avert holes being dug for the vines in the middle of his established beds; prospects for the future of the garden looked bleak until Lane was hastily replaced by yet another governor, Colonel Beatson. Burchell's elation at this 'glorious news' proved short-lived; even the pleasure of his official appointment as the Company's Naturalist at this time was soon undermined when the paucity of his proposed salary was revealed. The initial prospect of support from Beatson – who relayed to him the Company's encouragement for his botanical endeavours – seems to have soured within a short space of time and for some reason Beatson, remembered as one of the more enterprising projectors of improvement on the island, excluded Burchell from his plans (and literally wrote him out of the island's history which he compiled).[80] On 26 October 1808 Burchell recorded that

> the garden is fast falling to ruin owing to the little notice and support which it receives from the present Government, and notwithstanding, I represented ... to Col. Beatson the mortification I felt at thus witnessing all my labours fall to the ground merely from neglect of giving it from the Government that due support which I had all along hoped it would receive. Yet still nothing is done to remedy the evil.

On 5 June 1809 he proposed that 'as the Botanic Garden could never come to any good in the Valley, it would be best to abandon it altogether as soon as another

---

[79] Burchell's stay in St Helena and his achievements there are conveniently summarized in Cronk 1988.
[80] Beatson had been greatly displeased by the negative tone of a report prepared by Burchell for the Court of Directors of the EIC: Burchell declined Beatson's request that he should withdraw it, and from that point onwards their relationship became embittered. McKay (1934, 487) reproduces a report from Beatson to the directors expressing the conviction that all Burchell's efforts as naturalist 'should tend, as much as possible, to the immediate augmentation of food for man and beast ... All other projects are of secondary consideration.'

place could be found whereto the plants could be removed'. The governor declared himself of the same opinion, 'and at my request ordered that for securing the plants for the present, a rail should be run along before the house to entirely enclose it'. No more is heard of the scheme,[81] however, and soon afterwards Burchell resigned from his post.

It should be mentioned that he had also turned his attention to further aspects of the island's natural history during this time. Beatson had called on 28 July 1808, along with the lieutenant-governor and the chaplain, to see Burchell's 'St Helena mineralogy collection', on which occasion the governor recommended that he should be communicating his findings to the Royal Society. A particular point of interest was the prospect that some of the island's brightly hued clays could prove of artistic and commercial interest as artist's colours. When he mentioned this to T. H. Brooke (nephew of Governor Brooke and sometime acting governor of the island) on 2 April 1807, Brooke responded that his observations 'exactly confirmed what Mr Daniel, the Artist, remarked when they rode round the Island once together, when Mr D. frequently got down to pick up earths which he said would make excellent paints'.[82] A report was prepared for forwarding by the governor to the Court of Directors, together with 'a card with specimens of colours painted with balsam', and on 22 June 1807 '6 Boxes of St Helena colours of No. 1, 3, 4, 5, 6 and 8' were sent. Burchell later confided to Captain Halkett, homeward bound with the China fleet, that his 'greatest hope' rested with these colours. The directors proved encouraging and eventually clays of eleven individually numbered colours were forwarded to them; while Burchell had characterized them as potentially of interest to artists and interior decorators, we learn only that having been examined by Sir Joseph Banks they were deemed not suitable for the manufacture of china, after which no more is heard of them.[83] Lichens and marine algae were also collected as possible sources for dyes. At Burchell's behest, guano from sea-fowls was collected from Egg Island in the hope that it might prove commercially attractive as a fertilizer, but this too proved a false hope.

For all his admirable character and sentiments, Burchell's five-year sojourn on the island failed to deliver long-term satisfaction. On leaving for the Cape on 26 November 1810 he closed his journal by observing that 'it will appear to

---

[81] Burchell's creation seems not long to have outlasted his presence: in 1813 Dr Andrew Berry (previously involved in the project to grow nopal (prickly pear) at Madras: see MacGregor 2018, 70–71) advised Beatson that 'A portion of the Garden be devoted to the growth of the Cactus of most value for rearing the fine cochineal, as its objects as a Botanic Garden have been frustrated.' Later in the century the site was handed over to the War Office and in 1900 the last of the trees were cleared to make way for an extension to the barracks.

[82] Thomas Daniel (1749–1840) and his nephew William (1769–1837) must have called at the island on their way home following their seven-year sojourn in India, where they had built a considerable reputation as topographical artists and printmakers.

[83] McKay 1934, 484. Kitching (1937, 3) mentions that a further consignment sent at a later date likewise proved to be of no commercial value.

whoever reads it that my life at St Helena has been one uninterrupted scene of disappointments'. A major contributory factor had undoubtedly been the failure of his betrothed to join him (she married instead the captain of the vessel carrying her there), but it does seem doubtful that life would have turned out very differently had the marriage taken place. The major part of his career as a naturalist still lay ahead, but St Helena had brought only a glimpse of the success he would yet enjoy. His survey work at the Cape, untrammelled by authority (for having been frustrated in his ambitions once, he declined the post of Botanist to the Cape Colony when it was offered to him), would prove infinitely more rewarding: Malgosia Nowak-Kemp records that he returned from his five-year stint at the Cape – so much more rewarding than his similar time on St Helena – with a collection of 63,000 natural history specimens.[84]

Although undoubtedly he must have had ambitions to publish his observations on St Helena, on his return to London from the Cape (calling briefly at the island on the way) it would be to the publication of his *Travels in the Interior of Southern Africa* (1822–24) that his efforts would be devoted before setting out again to spend a further five years exploring in Brazil. The bulk of his collections remained with Burchell until his suicide in London in 1863, when his sister arranged the transfer of the botanical material to Kew. Also included in the gift to Kew were Burchell's manuscript 'Flora Insulae Sanctae Helenae', listing 200 plants in tabular form,[85] and a 'Flora Heleniana' containing manuscript descriptions of sixty-six of those plants; a further list of 475 plants includes species found by Burchell growing in gardens on the island. A folio volume of topographical and botanical sketches was also included in the gift.[86]

Quentin Cronk, the most assiduous chronicler of the island's botany, concluded that the vegetation as recorded by Burchell is essentially the same as that found in the island today (although many more species have been introduced into the wild in the interim, the 200 wild plants recorded by Burchell having now increased to some 350); the clear implication is that the major changes wrought on the island's flora had effectively been completed by the turn of the nineteenth century.[87] Even

---

[84] Nowak-Kemp 2018, 545.
[85] Cronk (1988, 49) mentions that part of a duplicate collection of plants sent home from St Helena in 1810 was seized by Customs and Excise and later auctioned by them; they were bought by a Mr A. B. Lambert, and it was only at the auction following Lambert's death in 1842 that Burchell was able to recover them, at a cost of 5 guineas. A complete transcription of the 'Flora' is provided by Cronk in an appendix to his paper.
[86] Cronk (1988, 49) mentions that a further volume of sixty-five sketches is now in the Africana Museum of the Johannesburg Public Library. Many of Burchell's botanical specimens had in the meantime been shared with (some delivered personally) the Swiss naturalist A. P. de Candolle, who published them in his compendious *Prodromus systematis naturalis regni vegetabilis* (1823–73).
[87] The earlier history of the introduction and evolution of St Helena's endemic species is elucidated in a detailed chapter by Ashmole and Ashmole (2000, 65–80), making extensive use of the work of Cronk and other naturalists.

Burchell's list includes plants that were certainly introduced, for the number of truly indigenous species that had survived to his day is now thought to have stood at no more than eighty.

WILLIAM ROXBURGH

One of Burchell's most assiduous correspondents and benefactors had been William Roxburgh (1751–1815), superintendent of the botanic garden in Calcutta and a major figure in his field (Figure 20b) to whom the epithet 'the father of Indian botany' has often been aptly applied. The relationship was undoubtedly a deferential one on Burchell's part: they must surely have met when Roxburgh passed through the island in 1805, but Burchell had already left for the Cape when Roxburgh, by now sixty-two years of age and with thirty-seven years of EIC service, paid a second and more protracted visit to the island in 1813. His health now in decline, Roxburgh had requested a leave of absence to visit the Cape or St Helena, and eventually England, 'should it be necessary' for his recuperation. Along with his request for leave on that occasion he had written as follows to the government in Calcutta:

> ... in consequence of repeated applications from the various Governors of St Helena for many years past and of orders from the Honorable the Court of Directors, plants, seeds, and roots of the most useful kind have been frequently sent from hence to that Island, for its improvement, but hitherto with much less success than was hoped for. Forest trees they want most of all, and I understand they have in general failed in rearing such trees from seed. Permit me therefore to request ... that it might be advisable to send by each of the companies ships which form the next Fleet, one or two Chests filled with growing plants of all those trees as appear to promise most success. Such as Sissoo Buddam Lundry, Teak, Sumatra Cassia, &ca &ca which the state of the nurseries in this garden can at present furnish. If this idea meet with the approbation of Government have the goodness to oblige me with the order for getting them ready as soon as you can. It is probable that I shall be at that Island at the time the plants will arrive, and in that case will with pleasure give the best advice and directions in my power for their planting and future management.[88]

The keen appreciation shown here for St Helena's desperate need of repopulation with forest trees is perhaps unsurprising in a curator whose garden had played a major part in the redistribution of plants throughout India and the wider empire. At a more significant level, Roxburgh was evidently aware of the consequences of the environmental damage wrought by French settlement on Mauritius, as observed by Pierre Poivre, a pioneer figure in identifying anthropogenic change as a major factor in landscape degeneration. This awareness on Roxburgh's part, supported by similar observations made by Robert Kyd in Calcutta and by

---

[88] IOR P/8/13; reproduced in Robinson 2008, 106.

Burchell on St Helena, led to the early emergence of the island as a site at which the environmental consequences of thoughtless destruction of forest and other vegetational cover – to the extent of the extinction of endemic plants – were exposed and where measures to counteract them were put to the test.

Roxburgh arrived on 7 June 1813 on the East Indiaman *Castle Huntly*; his journey onwards was dependent partly on his health and partly on the availability of a passage on one of the ships calling at the island on the homeward voyage.[89] A letter to Banks, written soon after his arrival on St Helena, concludes with the statement that, 'to please myself & Colonel Beatson, I shall begin upon the plants of this Island as soon as the Fleet has sailed'.[90] In the end his lengthy sojourn on the island, extending to nine months, must have transformed the nature of that undertaking; certainly it was sufficient for Roxburgh, one of the Company's most industrious botanists, to produce an annotated list of the island's flora, although the combination of poor health and the island's demanding topography prevented it from being exhaustive in scope.[91]

Nonetheless, Roxburgh's *Flora Sta. Helenica*[92] lists 363 species of flowering plants and twenty-five ferns (including nearly twenty species described there for the first time); thirty-three of them were believed to be indigenous to the island, although most of the new names coined by Roxburgh have since been superseded.[93] A considerable proportion of the plants he encountered had already been noted by Burchell, but since the latter's manuscript *Flora* remained unpublished he lost precedence in terms of nomenclature. For the researchers who followed in their footsteps – notably Darwin – it was therefore Roxburgh's list (particularly in the form published by Beatson) that proved particularly influential.

Charles Darwin

The endemic status of part of St Helena's flora was well established by the time Charles Darwin (1809–82) spent six days on the island in 1836, while traversing the South Atlantic on the *Beagle*: Grove has noted that Darwin could state that the island possessed 'an entirely unique flora' without himself having carried out

[89] Roxburgh was accompanied throughout the voyage home by his wife and three children. On reaching London they took up residence in Cheyne Walk, Chelsea, but soon moved to Edinburgh where he died.

[90] Robinson 2008, 74. Beatson and Roxburgh had met in Madras twenty years earlier, when the latter was involved in a canalization scheme in the Northern Circars.

[91] The 'Alphabetical List of Plants seen by Dr Roxburgh growing on the Island of St Helena, in 1813–14', produced as an appendix to Beatson's *Tracts relative to the Island of St Helena* (pp. 295–326), acknowledges its incomplete nature, for 'Dr Roxburgh's bad state of health during his residence here … did not admit of his undertaking such a work'.

[92] The *Flora* was first published under that title (though anonymously) in St Helena in 1825, although its contents had already been published by Beatson (see above). See Cronk 1987 and 1989 for further discussion of the authorship of the list.

[93] Robinson 2008, 105–06. See additionally Turner 2016.

the survey work to support such an assertion. There can be no doubt that it was Roxburgh's extensive list of plants – either in the form published by Beatson in 1816 or in the anonymous monograph of 1825 – that provided him with this insight.[94]

More than any of those who had preceded him, Darwin (Figure 20c) showed an awareness that from this 'little world, within itself', not only had many species been lost but that prominent among the reasons for these extinctions had been the eagerness of the settlers (and indeed of the directors of the EIC who drove many of the initiatives) to populate the landscape with potentially more productive alien plants. The predominantly English character of the vegetation now established on the island struck Darwin forcibly, causing him to reach some uncomfortable conclusions as to the environmental consequences:

> In latitude 16° & at the trifling elevation of 1500 ft, it is surprising to behold a vegetation possessing a decided English character. But such is the case; the hills are crowned with irregular plantations of scotch firs; the sloping banks are thickly scattered over with thickets of gorze, covered with its bright yellow flowers; along the course of the rivulets weeping willows are common, & the hedges are formed of the blackberry, producing its well known fruit. When we consider the proportional numbers of indigenous plants being 52, to 424 imported species, of which latter so many come from England, we see the cause of this resemblance in character. These numerous species, which have been so recently introduced, can hardly have failed to have destroyed some of the native kinds … Many English plants appear to flourish here better than in their native country; some also from the opposite quarter of Australia succeed remarkably well, & it is only on the highest & steep mountain crests where the native Flora is predominant.

These bleak conclusions on the island's flora provided Darwin with persuasive data on the questions of endemism and extinctions. He wrote, for example, that:

> In St Helena there is reason to believe that the naturalised plants and animals have nearly or quite exterminated many native productions. He who admits the doctrine of the creation of each separate species, will have to admit that a sufficient number of the best adapted plants and animals have not been created on oceanic islands; for man has unintentionally stocked them from various sources far more fully and perfectly than has nature.[95]

It is well known that the Galapagos Islands had provided Darwin with key data in the development of his ideas on the concept of natural selection, but rather than springing fully formed on the spot these inklings were painfully worked out by a process of sifting and refinement that began on the long circumnavigation that carried the *Beagle* across the Pacific and round the Cape of Good Hope. These

---

[94] Grove 1995, 355.
[95] Darwin 1859, 389.

extracts confirm that by the time he left St Helena in July 1836 they had already begun to coalesce into the conclusion that Darwin himself proved so reluctant to acknowledge: that 'the doctrine of the creation of each separate species' had run its course and that the evidence provided by the natural world itself carried irrefutable evidence for the processes of evolution – and extinction. In particular, he was intrigued by the implications of some layers of fossil shells he investigated, in which he recognized certain land snails belonging to species that were now extinct. In this way, as Richard Grove observes,

> St Helena provided Darwin with some of his best data on the dynamics of island populations, endemism and extinctions. Thus in the *Origin of Species* St Helena occupies a prominent place in the section on the inhabitants of oceanic islands, actually being mentioned, in comparison with Ascension Island, much earlier in the text than the Galapagos Islands.[96]

Other features of the island's physical and natural history also caught Darwin's imagination, not least its volcanic structure, which he well recognized while misinterpreting some details of its formation.[97]

Joseph Dalton Hooker

St Helena was a comparatively accessible and unchallenging way-station for the Antarctic expedition mounted in 1839–43 under the command of Captain (later Sir) James Clark Ross in the Royal Navy ships *Erebus* and *Terror*. The primary task of the expedition was to further the earlier investigations on the subject of terrestrial magnetism, including the establishment of magnetic observatories at St Helena and at the Cape as mentioned above. Joseph Dalton Hooker (1817–1911), who sailed as assistant surgeon and naturalist on HMS *Erebus*, had additional instructions from the Royal Society relating specifically to recording what survived of the island's endemic flora – an interest already sparked by his friendship and collaboration with Darwin.

Hooker (Figure 20d) was very much aware of the critical state in which the island's indigenous flora now found itself – already decimated by human activity and by the introduction of grazing animals and under continuing threat from the advance of aggressive alien plant species that threatened to stifle the surviving

---

[96] Grove 1995, 362–63. Here Grove makes it plain that by reference to Burchell's and Roxburgh's earlier lists of plants, Darwin 'was able (like Burchell) to associate the rapid speed of environmental change on the island, in both historical and geological time scales, with changes wrought on the endemic flora by deforestation and by the invasion of alien species, "which can hardly have failed to destroy some of the native kinds"'.

[97] As an interesting aside, after his return to England Darwin visited the EIC's military seminary at Addiscombe in order to check certain details of the volcanic topography against the large-scale model of the island kept there, constructed by Robert Francis Seale, author of an early book on the island's geology (see Chapter 5); see also https://www.darwinproject.ac.uk/letter/?docId=letters/DCP-LETT-427.

endemic flora. By combining the indigenous species identified by Roxburgh with those of his own observation, he concluded that they now reached a total of about forty-five, compared with 110 naturalized plants – a telling illustration of the advanced state of ecological change.[98] Following his later appointment as director of the Royal Botanic Gardens at Kew, although he promoted the diversification of horticulture on the island, Hooker also acknowledged the need to protect the threatened indigenous flora: 'It is most desirable also', he wrote, 'that some of the extremely interesting native plants, which have become very scarce, should be preserved from extinction'. At his behest, seeds and plant specimens of endangered species were gathered on the island and transmitted to Kew for safe preservation.

Hooker was also active in introducing new economic species to the island in an attempt to render it more self-sufficient. He continued the work of his father (who had preceded him in office at Kew) in promoting the transfer there of cinchona, an important source of anti-malarial quinine – perhaps the most important drug in the Indian surgeon's medicine chest. Seeds collected in South America were germinated in a purpose-built glasshouse at Kew, funded by the India Office, before being shipped to St Helena as well as directly to Ceylon and India.

John Charles Melliss

The life of one historic contributor to science on St Helena followed a trajectory counter to those considered so far, for having been born and lived on the island until the age of thirty-five, J. C. Melliss (1835–1910) had by that time completed his work in the field and left for England; he would, however, remain engaged with island matters for much of the remainder of his life. His father, Captain G. W. Melliss of the St Helena Artillery, is particularly remembered for having been responsible for the construction of the inclined plane on Ladder Hill, completed in 1829; his career with the East India Company ended with the transfer of the island's administration to the Crown in 1834, but Melliss was one of the fortunate few to have been re-employed by the colonial service, first as Surveyor and later as Civil Engineer.[99] J. C. Melliss, having studied at King's College London, served in the Royal Engineers and worked for the Metropolitan Board of Works, returned to the island as Clerk of Works before following his father in the role of Civil Engineer.

The younger Melliss's early interest in natural history studies had been encouraged by both his parents, progressing to the point where he was in correspondence with several of the foremost natural scientists of his day. He also supplied them with specimens: one recipient, the Revd O. Pickard-Cambridge – the foremost specialist on spiders of his day – responded by naming one species,

---

[98] See Desmond 1999. Hooker was able to discern the African origins of some of the indigenous plants and the South American affinities of others, including the cabbage trees (ibid., 33). See also Grove 1995, 364.

[99] The information here was gathered by Trevor Hearl and reproduced in Ashmole and Ashmole 2000, 83.

Argyrodes mellissi, the Golden Sail Spider, in Melliss's honour.[100] It came as a loss to the whole island when, in a round of government retrenchment in 1870, Mellis's post was abolished and he was forced to find employment in England; after overseeing several large-scale hydrological projects in the Midlands, he founded his own company, which survives today as Melliss LLP, civil and structural engineering consultants.

Among the island's natural resources, Melliss's particular engagement lay with its botany. Although he authored some 'Notes on the birds of the island of St Helena', his opening remark that 'The feathered portion of the St-Helenian fauna can scarcely be said to be so interesting from a scientific point of view as the rest'[101] reveals the nature of his primary interests.

Within his busy work schedule in England, Melliss found time to consolidate the results of his earlier researches, which he published in 1875 as the compendious *St Helena: A physical, historical, and topographical description of the island, including its geology, fauna, flora, and meteorology* (Plate 20). Fellowships of the Linnean and the Geological Societies acknowledged the high quality of his work; the plant genus *Mellissia* was named in his honour by J. D. Hooker, and the centenary of the volume's appearance was marked by the issue in St Helena of a set of postage stamps based on his illustrations.

Although he published nothing further in this vein, Mellis remained concerned for the wellbeing of the island: in 1906 he lectured to the Royal Colonial Institute on the consequences of the withdrawal of the island's garrison, and two years later he was a founding member of the St Helena Relief Committee.

More recent scientific investigation

The later twentieth and the twenty-first centuries have seen an enormous increase in scientific interest in the island as a laboratory for the investigation of a diversity of phenomena. The personnel involved are too numerous to review in detail, but by way of illustration two major contributions may be mentioned. Philip and Myrtle Ashmole's lengthy fieldwork resulted in their *St Helena and Ascension Island: A Natural History* (2000), which consolidated and expanded the record of the island's ecological history, providing insights from the earliest millennia of the population of the sterile volcanic landmass to the identification of hitherto unrecorded species of the present day and the perilously balanced survival of other organisms. Quentin Cronk, now Professor of Botany at the University of British Columbia, who has dedicated a great deal of attention to the history of the island's vegetation, published *The Endemic Flora of St Helena* (2000), in which he documented the destruction of the island's fragile ecosystems, and has contributed to strategies for preserving and recovering its much-diminished plant cover.

---

[100] Ashmole and Ashmole 2000, 82.
[101] Mellis 1870, 97–98.

Scientists in transit: St Helena as a site for scientific investigation

A feature of progress in more recent years, however, has been the way that so much important fieldwork has come to be undertaken by the island's own residents. Recent investigators have taken particular inspiration from the example of the brothers Charles and George Benjamin. George Benjamin (1935–2012), a forest guard with the Agriculture and Forestry Department, worked closely with Quentin Cronk and on one of their outings in November 1980 spotted a curious flowering plant clinging to a dangerously precipitous cliff face near a spot called the Asses Ears: neither felt up to scaling the cliff, some hundreds of feet above the sea, but two days later they returned with Charlie Benjamin (1932–2007), a fisherman used to clambering the cliff faces to reach promising spots for a catch; Charlie climbed down with no more than a rope around his waist and returned with some cuttings in a bag (and a flowering stem between his teeth) to reveal an incredibly valuable prize – a specimen of St Helena Ebony, long thought to be extinct. From the seeds and cuttings derived from the plant,[102] George began a one-man propagating mission that saw the species successfully re-established in selected wild spots, while back in the UK Cronk was able to use the discovery to add impetus to his efforts to muster support for a campaign of botanical conservation on the island. The present-day inheritance of those initiatives – again involving much local input as well as scientific backing from university and government agencies – is further discussed in Chapter 9.

---

[102] Charlie returned a second time to sample an adjacent plant and retrieve some soil samples, but understandably declined to make a third descent.

7

Napoleon on St Helena

Napoleon Bonaparte's dazzling military adventures in the European theatre seemed to have run their course with his defeat by the Allied forces and his abdication on 6 April 1814, followed by incarceration on the isle of Elba. It was perhaps characteristic of him that in less than a year the emperor's star once more flared into life with his escape from imprisonment and his dramatic (though short-lived) re-entry to the Tuileries Palace, but the prospect of his renewed domination of the Continent was finally extinguished with the crushing defeat at Waterloo. In the aftermath of the battle, Napoleon himself made hastily for Paris and then for Rochefort on the Atlantic coast, where a rendezvous had been arranged with a frigate that might carry him to freedom in America. An impregnable blockade by the Royal Navy put that prospect out of reach and – with Bourbon and Prussian forces closing in on him – on 15 July the fugitive emperor presented himself on board HMS *Bellerophon*, where he announced to the commander, Captain Frederick Maitland, that he had come to place himself 'under the protection of the laws of England'.[1] The courtesies with which the formalities were carried out were indeed more appropriate to a noble seeker of asylum rather than a prisoner of war:[2] Maitland gave up his own cabin to Napoleon's use, while the members of his accompanying entourage were accommodated as best they could be; after a short delay, *Bellerophon*, with her extraordinary French complement, set sail for England, with an air of measured politeness prevailing on all sides.

At this point Napoleon evidently had every hope that his days might be lived out in congenial retirement in the comfort of a secluded country estate (preferably

---

[1] For Captain Maitland's firsthand account of these intricate proceedings, see Maitland 1904. Every word of his carefully constructed narrative was undoubtedly weighed with a view to substantiating his actions in the eyes of his naval superiors. The virulently dissenting interpretations placed by French commentators on the conduct of these proceedings have scarcely dimmed in the past two centuries: see, most comprehensively, Martineau 1971.

[2] If Napoleon's expectations seem optimistic, it may be noted that Barry O'Meara (1822, I, 84), who was to accompany him as surgeon to St Helena, records being told personally by Bonaparte that, 'Before I went to Elba, Lord Castlereagh offered me asylum in England, and said, that I should be very well treated there, and much better off than at Elba', an offer which, although declined, 'had much influence with me afterwards'.

Napoleon on St Helena

'ten or twelve leagues from London'). But in Whitehall the cabinet had already discussed his fate in the event of his recapture, with a primary concern that there should be no opportunity whatever for a repeat of the Elba fiasco: even before *Bellerophon* had left Rochefort, the prime minister of the day, Lord Liverpool, had informed his foreign secretary (then in Vienna) that it was the collective resolve of the cabinet that there could be no question of confinement on English soil. Fearing that the vanquished emperor would become 'the object of curiosity'[3] and 'possibly of compassion in a few months' – not to mention the danger that he would act as a focus for continuing ferment in France – the government had concluded that St Helena would be 'the place in the world best calculated for the confinement of such a person'.[4] The island had several advantages:

> the situation is particularly healthy ... There is only one place ... where ships can anchor, and we have the power of excluding neutral ships altogether ... At such a place and such a distance all intrigue would be impossible; and, being so far from the European world, [Napoleon] would soon be forgotten.

Few prophecies can ever have been so comprehensively unfulfilled.

The tenor of the shipboard formalities changed with the arrival of an emissary from Whitehall, under-secretary of state Major-General Sir Henry Bunbury, to inform 'General Bonaparte' – the only title the British would allow him – of his exile.[5] Following the – predictably violent – protests, hotly expressed by all the French party and with Napoleon still asserting in a final, futile appeal addressed to the Prince Regent ('the most powerful, the most constant, and the most generous of my enemies') that he had voluntarily come on board, that he was indeed 'the guest of England',[6] on 8 August, together with a retinue of twenty-five, the disconsolate

---

[3] As indeed witnessed when the ship touched at Torbay and at Plymouth, where great flotillas of small boats came out in the hope that the man whose very name had formerly struck terror along the south coast might be glimpsed in his captivity.

[4] Cipriani, the *maître d'hôtel* of the detainees' household, later claimed that while he had formerly been in exile on Elba the emperor had been told of a plan drawn up at the Congress of Vienna to send him to St Helena and that this news had indeed 'contributed to determine Napoleon to attempt the recovery of his throne' (O'Meara 1822, I, 84).

[5] The form of address was to be a continuing source of bitter dispute. While some claim that the British government had never recognized Napoleon as emperor of the French, Michel Dancoisne-Martineau has pointed out to me the falsity of this claim and the fact that he was widely acknowledged in England by that title; moreover, by the time of the ratification of the Treaty of Fontainebleau (April 1814), the 'imperial titles' had been written into the British official papers. The government, however, determined that he should now be addressed as and treated with the deference due to a general on the retired list. Throughout the period of his incarceration, neither side budged an inch in acknowledgement of the other's claim.

[6] Forsyth 1853, I, 10–11. The distinction is one that continues to divide commentators largely along national lines, although there were nonetheless contemporary supporters in England of Napoleon's viewpoint, most notably in the person of Lord Holland,

ex-emperor transferred to HMS *Northumberland*. There was no longer room for doubt as to his status: henceforth he would be 'prisoner of State of His Britannic Majesty'. Under the command of Captain Charles Ross and flying the flag of Rear-Admiral Sir George Cockburn, responsible for safely conveying the general to St Helena, the *Northumberland*, accompanied by a small escort squadron, set out on a voyage that would culminate sixty-seven days later, on 17 October, in the roads off Jamestown.

News of the impending arrival of the scourge of Europe, now destined for indefinite detention on this 'speck in the ocean', had reached St Helena only four days before the arrival of the *Northumberland*, sparking some frantic logistical activity. Security being a first priority, the government sent with him from England a battalion each from the 66th (Berkshire) Regiment and the 53rd (Shropshire) Regiment – a total of some 1,300 men; a second battalion of the 66th arrived in 1816, the 53rd was recalled to England the following year and the 20th (East Devonshire) Regiment arrived in 1819. A further 900 men from the St Helena Regiment and the St Helena Artillery were mobilized to man the gun emplacements at strategic points around the island and to mount pickets, while a detachment of twenty dragoons was transferred from the Cape to facilitate communication with the governor. At sea the Royal Navy stationed HMS *Newcastle* to guard the approaches to Jamestown while a patrolling force comprising a frigate and two brigs circled the island on a regular basis; Ascension Island – 800 miles away – was garrisoned for the first time as an additional measure against seaborne rescue attempts. Together with the ancillary military, naval and civilian staff (in some cases accompanied by their wives), the influx doubled the population of St Helena at a stroke, provoking immediate crises due to shortages of food, water and accommodation.

In addition to Napoleon himself, the French contingent comprised the following, accompanying him voluntarily into exile: Count and Countess Bertrand and their three children; Count and Countess Montholon and their only child; Count de Las Cases and his teenaged son; General Gourgaud; four *valets-de-chambre*, two grooms, a footman, *maître d'hôtel*, cook, butler, steward and two servants; the French surgeon who had accompanied Napoleon as far as Plymouth had declined to go further and at Napoleon's instigation the Irishman Barry O'Meara, surgeon of the *Bellerophon*, had been chosen as a substitute, an appointment approved by the Admiralty.

The pragmatic Admiral Cockburn had been instructed that on arrival at St Helena the French party should remain on board until such time as suitable accommodation could be found for them, but having shared in the crowded crossing he decided instead to allow them to disembark into temporary housing on the evening of their arrival. The former emperor and his immediate entourage were to be put up in a house which lay towards the bottom of the town, belonging

---

who spoke passionately in his favour in the House of Lords. See Martineau 1968 for a full account of the proceedings.

to Henry Porteous, superintendent of the EIC's gardens in Jamestown. Although considered today to have been 'the finest house in Main Street', Napoleon took violently against these modest lodgings (which lacked so much as a garden or courtyard and was all too open to the gaze of the crowds that had pressed around to witness his disembarkation). An alternative had to be found after a single night.[7]

The question of longer-term accommodation was settled the following morning when Admiral Cockburn and Governor Mark Wilks conducted Napoleon on a tour of inspection of Longwood House, a disused and crumbling single-storey building – a former barn that had been repurposed as a summer residence for the lieutenant-governor of the island. Although all too obviously in need of repair and considerable extension to render it serviceable, Longwood was in fact deemed the only house on the island that might hold them (not counting the governor's own residence, which was specifically excluded from consideration).[8] Napoleon was distinctly unimpressed, but in the absence of an alternative it was decided that the necessary works to make it habitable should be put in progress.

For more immediate purposes, the choice now fell on The Briars, an attractive bungalow which caught Napoleon's eye on the outing (Plate 21). At a small distance from Jamestown, this was the home of the EICs Superintendent of Public Sales, William Balcombe, and his wife and five children. Accommodation was duly found there for Napoleon in a summerhouse or pavilion in the garden, where he would work, eat and sleep on the campaign bed he favoured, all in a single room; Las Cases and his son occupied a small chamber above while other members of the party were housed nearby. Additional space was later provided by a marquee, erected near the pavilion. The gardens at The Briars extended over several acres, offering pleasant shady walks among the fruit trees and flowerbeds. At first sentries were posted at the house, but soon they were withdrawn on the orders of Admiral Cockburn (by now in command of military matters on the island), who considered them an unnecessary intrusion. The general and his entourage were allowed to roam at will not only within the garden but throughout the island, with only the fortifications being out of bounds.

For seven weeks Napoleon and the Balcombe family lived amicably enough alongside each other, playing chess and cards together and enjoying a remarkably easy relationship; in particular, the Balcombes' second daughter, fourteen-year-old daughter Betsy, provided congenial company for the general and brought out an unexpectedly gentle, avuncular side to his character.

---

[7] The house had earlier provided adequate accommodation for Napoleon's later adversary, Colonel Sir Arthur Wellesley (later Duke of Wellington), when he passed through the island on his return to England from service in India.

[8] Michel Dancoisne-Martineau has observed that this claim was a fiction: the properties at Rosemary Hall, Oaklands, Oakbank and Farm Lodge were all infinitely superior, he suggests, especially in terms of climate; while Longwood may have been recommended for strategic purposes, one is left to wonder whether vengefulness might not have played some part in the choice.

By the time initial work had been completed at Longwood, allowing the new inhabitants to move there on 10 December, it had been made habitable (though little more). The exigencies of life on St Helena combined to render the reconstruction process a laborious one: all the timber (and even building stone) required – not to mention the furnishings – had to be carried up to the site, a task that employed 200 to 300 seamen on a daily basis, aided by fatigue parties from the 53rd Regiment. Building work (mostly in timber) was carried out by ships' carpenters from the squadron, aided by a number of tradesmen from the island (the latter evidently astonished by the speed at which the sailors worked). In the space of six weeks the original four or five rooms of the modest house (themselves the product of haphazard extensions added as required by earlier occupants) had been enlarged to accommodate the whole party, apart from the Bertrands who occupied a small house a mile away at Hutt's Gate until new premises could be constructed for them. A print executed within a few decades of Napoleon's death (Plate 23) shows the house at its fullest extent (allegedly thirty-six rooms in all), though this was by no means achieved immediately: repairs and extensions would continue throughout the period of its occupation – much to the chagrin of its principal occupant.[9] Ultimately, a salon and dining room, library and billiard room were included, as well as the necessary sleeping quarters (those of the servants being in the attics).

It should be mentioned here that quite another house had been envisaged by the British government for the accommodation of the French. Designs for an entirely new structure were drawn up, an extensive suite of furniture was commissioned and the fittings and decorations – all specified to a high degree – were procured in London.[10] New Longwood House, as it was to be known, ultimately had a building history as long as Napoleon's own incarceration: situated at a short distance from the original building, work on it was eventually begun three years after Napoleon's arrival on the island and was completed in January 1821 (according to entirely different designs drawn up on the island by Major Anthony Emmett of the Royal Engineers), some three months before the death of its intended occupant. Napoleon himself remained equivocal about accepting the long-term implications embodied in a permanent residence: he made only one recorded visit to the site while it was under construction and in the end declined to move due to failing health.[11]

[9] An undated 'Statement shewing the probable annual expenditure on account of General Bonaparte', reproduced by O'Meara (1822, II, 450–51), includes an entry for 'public mechanics employed at Longwood House, whose services are likely to be wanted for a considerable time': it includes '2 overseers, 6 carpenters, 4 sawyers, 5 masons, 3 plasterers, and 1 painter'.

[10] On 25 October 1815 the *Morning Chronicle* carried an account of the intended house, for which 'the framework … is nearly completed at Woolwich'. Measuring some 100 by 120 feet, with fittings and furnishings specified to the highest degree, the picture painted is of a fashionable residence quite unlike the quarters the general would occupy on the island.

[11] The history of New Longwood House and of the 2,000 tons of materials, furnishings and decorations commissioned and supplied for it (a great deal of which was

The broad plain on which Longwood House was set has been variously described, according to the sympathies of the particular commentator. For William Forsyth, author of the *History of the Captivity of Napoleon at St Helena* (1853), for example:

> The plains in the upland ... which exist only at Deadwood and Longwood, are by no means barren ... And the climate here is more healthy than in the valleys. The heat of the sun is tempered by a refreshing breeze, which, wafted from the southern ocean, envelops the more elevated parts of the island in a shroud of mist.[12]

Admiral Cockburn too found the area around Longwood 'perfectly adapted for horse exercise, or for pleasant walking, which is not to be met with in all the other parts of the island'. It was, he maintained, 'allowed to be beyond comparison the most pleasant as well as the most healthful spot of this most healthful island'. The French view could scarcely have been more at odds: for General Montholon, the circuit within which they were confined 'comprises only arid rocks and ravines which the eye cannot contemplate without horror'; the climate was the most uncongenial on the entire island – 'It is always windy, and it rains every day. We live in the midst of clouds and in a very damp atmosphere.' Within a short time of arrival there, he continued, 'The Emperor feels his health giving way, and we all suffer more or less.' The entire entourage claimed to have been afflicted with constant bronchial problems; even the playing cards that provided their principal recreation had to be regularly dried out in an oven. The place was also infested with mosquitoes, cockroaches and rodents, of which the last posed a particular problem: 'a great quantity of linen and other effects has been rendered useless by the rats, and this for want of closets'.[13]

The War Department had decreed that 'no greater measure of severity, with respect to confinement or restriction, be imposed than what is deemed necessary for ... the perfect security of General Bonaparte's person'.[14] Had they known of such an order, the French would certainly have protested at its interpretation on the island following the arrival of a new governor in April 1816, a move that brought to a close the comparatively benign regime administered jointly by retiring civil governor Mark Wilks and Admiral Cockburn, in charge of the military. The new man (in whom both positions were combined), General

---

diverted for more immediate use in the existing house, where initially 'the few articles of furniture in [the] apartments had evidently been obtained from inhabitants of the island') is told in a commendably comprehensive and fully illustrated article by Martin Levy (1998).

[12] Forsyth 1853, I, 28.
[13] Ibid., I, 215; *Napoleon's Appeal* 1817. O'Meara (1822, I, 494) confirms the presence at Longwood of rats 'in numbers almost incredible', which he has seen 'assemble like chickens round the offal thrown out of the kitchen'; he also gives an account of the periodic attempts to exterminate them with the aid of dogs and clubs.
[14] Forsyth 1853, I, 15.

Fig. 22   General Sir Hudson Lowe (1769–1844), Napoleon's unbending gaoler on St Helena.

Sir Hudson Lowe (Figure 22), had been hand-picked in Whitehall to oversee Napoleon's detention; he arrived on the frigate *Phaeton* with a reputation as a 'safe pair of hands' – someone who could be relied upon to follow to the letter the detailed regulations that had been drawn up in London for the ordering of his charge's life. Posterity has not been kind to Lowe: his dogged assiduousness in the narrow interpretation of those orders has been widely characterized as – at best – unimaginative pedantry, but more widely as petty-mindedness if not downright vindictiveness.[15] Only in recent years has a more balanced assessment

---

[15]   Frank Giles, himself a former Whitehall mandarin with an acute insight into relationships within the corridors of power, gives perhaps the most balanced account of the Napoleonic

of his character emerged, with recognition that his appointment placed him in an unenviable position – as one of his French charges put it, between the hammer of Whitehall and the anvil of Napoleon.

In the course of his army career Lowe had been a participant in and witness to a number of setbacks that had befallen Napoleon's campaigns; he had even commanded the Corsican Brigade, a unit recruited from amongst those resistant to French rule (men characterized by Napoleon as 'vagabond Corsican deserters, Piedmontese and Neapolitan brigands'),[16] which oversaw the withdrawal of the French army from Egypt following its humiliation there. Few officers less congenial to the former emperor could have been found in the British army (as must have been evident to those who appointed him) and indeed the two men instantly developed a deep, mutual antipathy,[17] one determined to regulate the conduct of the French captives precisely as he had been instructed and the other intractable in refusing to acknowledge any such jurisdiction, either in principle or in the authoritarian detail in which it was applied.

To the new governor, it appeared that Admiral Cockburn had 'taken upon himself to grant much more indulgence, and a much greater space ... than he had any right to do'. No such leniency could be expected henceforth. As to General Bonaparte, Lowe was in no doubt of their relationship:

> He had better reflect on his situation, for it is in my power to render him much more uncomfortable than he is ... He is a prisoner of war, and I have a right to treat him according to his conduct.[18]

Clearly the niceties of Bonaparte's imperial status were of less concern to Lowe than the absolute demands of security: he immediately increased the guard at Longwood, stipulating 125 men for day-time guard duty and 72 at night – numbers viewed by Napoleon as provocatively oppressive and which, by any standards, seem excessive in this remote and inaccessible spot. Two perimeters were established about the house: within the inner one, about four miles in extent and with a sentry posted every fifty paces, Napoleon and his party were free to come and go as they pleased; within the outer perimeter, amounting to some twelve miles and taking in most of Longwood and Deadwood Plains, they could range unaccompanied on horseback

---

interlude on St Helena. Nonetheless, he finds Lowe 'an irascible, irritable, insecure and short-tempered man' who, while he found himself always under intense scrutiny from his masters, 'positively invited such attention by his unnecessarily frequent reporting, which betrayed his lack of confidence and anxiety not to put a foot wrong' (Giles 2001, 184).

[16] O'Meara 1822, I, 94. Lowe by no means merited the poor opinion of him fostered by Napoleon: Marshal Blücher and Sir John Moore both had reason to think highly of his service, though the Duke of Wellington was not an admirer.

[17] The gift of a gold watch, presented by Napoleon to Lowe and now in the National Army Museum (inv. no. 1963-10-212), must date from the very earliest days of their acquaintance.

[18] For Hudson Lowe's difficult interviews with Napoleon see especially Forsyth 1853, I, chapter IV.

or by carriage[19] but were liable to encounter many more troops – notably around the major encampment at Deadwood; the area also included a racecourse where islanders and the garrison met from time to time in competition.[20] The territory beyond those limits could equally be visited (with the continuing exception of the fortifications), but here the general had to be accompanied by the orderly officer of the day. Being unwilling to accept any kind of close watch placed on him when he went out of doors, Napoleon reacted by spending increasing amounts of time without stirring from the house. Objecting strongly to the curtailment of his visitors, he responded by cutting off all personal communication with the governor: after half a dozen encounters, the two men never again met face-to-face during the remaining five years in which Longwood would act as the former emperor's detention centre.[21]

The General – which title Lowe was punctilious in insisting upon – was subjected to a curfew (as indeed was the entire population of the island outside Jamestown) and he was heavily guarded: one battalion of troops was assigned to a camp established at Deadwood Plain – their tented lines all too visible across a ravine from Longwood; the other major encampment was at the head of the James Valley, above The Briars. Sentries were posted around the outer boundary of this area, preventing any further access towards the coast and its fortifications; at night the guard drew close around Longwood House, with two sentries at the front door; ditches 8 to 10 feet deep were dug to enclose the garden. The local population was excluded entirely from the area without specific authorization. A flag telegraph station established at Longwood constantly informed the governor in Jamestown of Napoleon's immediate circumstances: a series of prearranged signals (Plate 22) conveyed news varying from 'All is well with respect to General Bonaparte and family' to 'General Bonaparte is missing'![22] Twice daily a duty dragoon carried more detailed written reports from the orderly officer stationed within the household to the governor at Jamestown; each had to include confirmation that Bonaparte had been seen with the officer's own eyes.

True to character, Lowe proved an unbending gaoler. Following instructions he received from Whitehall, the running costs at Longwood were gradually reined in, especially in the provision of food and wine – even salt – allowed to the

---

[19] A carriage was originally lent for the general's use by Governor Wilks (Forsyth 1853, I, 35), but O'Meara (1822, I, 123) later mentions a phaeton, purchased with Napoleon's own money.

[20] One of the numerous infractions that brought Lowe's displeasure down on O'Meara involved the loan of one of Napoleon's horses to Betsy Balcombe for a race there.

[21] Three commissioners sent by the Russian, Austrian and restored French governments under the terms of the 1815 Paris Convention arrived on the island in June 1816; they received the same treatment, never succeeding in arranging a face-to-face meeting with Napoleon during the whole of their extended stay.

[22] Stirling University Library, MS11. The signals as illustrated form part of a codebook drawn up by Captain Henry Pritchard, in command of the garrison artillery, for Governor Wilks. The volume also contains instructions that should Napoleon go missing, 'Patroles' were to be 'sent out in every direction to insure the impracticability of any Person escaping from the Island'.

household;[23] coal and firewood were reduced, despite the inherent dampness of the site; newspapers were frequently held back and eventually stopped. Even water was in short supply (Napoleon's predilection for long baths became a source of irritation to Lowe when, as he put it, the garrison barely had water enough to cook their victuals); repeated requests were made for the provision of a water cart to provide for the household or for a conduit to be built to the house, the need for the latter becoming all the more acute when for a time Napoleon threw himself into gardening as a pastime – again with productive as well as aesthetic ends in mind.[24] (The enterprise was one that was viewed sympathetically – encouraged, even – by the British authorities, with the predictable exception of Hudson Lowe.)[25] In order to increase their personal spending power to mitigate these restrictions, items were periodically liquidated by the French: several groups of imperial plate were sold off (the arms and devices embossed on some pieces first having been erased).[26]

In the face of all these restrictions Napoleon began to languish, becoming lethargic and morose, neglecting even the forms of exercise that were urged on him by his physician. He suffered continually from headaches, swollen legs, bleeding gums and nausea. Much of his energy now went into dictating his memoirs; often days on end would be spent poring over journals and campaign maps with a view to compiling a narrative of his military exploits. This he dictated to his staff, taking every care in constructing his literary legacy to buttress the splendour of the military victories he had enjoyed and to burnish the glories that ultimately had eluded him and brought him to St Helena.

By 1820, it was clear that he had become seriously ill, suffering acute abdominal pains and diarrhoea. For a time he became convinced he was being poisoned: dictating his will in April 1821, he observed: 'My death is premature. I have been

[23] In addition to the costs of the consumable provisions, O'Meara (1822, II, 450–51) reproduces an (undated) estimate for 'public transport conveying the supplies furnished by the Purveyor to Longwood': it involved 'Forage for 8 mules daily; Pay of 2 muleteers in charge of the same; rations of ditto; pay of 2 soldiers ditto ditto' – and annual cost of £577 7s. 7d.
[24] Such a facility, it was suggested, would not only allow the household to support itself with fresh vegetables but could produce a surplus to the benefit of the garrison. Perhaps more importantly, it offered the possibility of growing trees and shrubs that would provide much-needed shade as well as privacy. To further divert the captive, General Bertrand designed an aviary, which he had constructed by a Chinese carpenter, but finding the plight of its inhabitants too similar to his own, Napoleon was inclined to liberate the birds. The original aviary survives today at the Musée Bertrand, Châteauroux. The whole enterprise has been magnificently reviewed by Donal McCracken (2022, esp. 136–54).
[25] Writing from the Colonial and War Office on 2 June 1820, Earl Bathurst assured the governor that if Napoleon were to express a desire for any plants from the Cape, the other British colonies or from Britain, 'no effort on my part shall be wanting to procure and forward them to St Helena in the manner best calculated to insure their safe arrival' (quoted in McCracken 2022, 152).
[26] O'Meara 1822, I, 122. The plate was deposited with the firm of Balcombe, Cole & Co. on the island. Balcombe – in whose house Napoleon had stayed earlier – is also identified as the 'purveyor' responsible for delivering provisions to Longwood; he was removed from his post by Lowe in 1818, on suspicion of having become too close to the general.

assassinated by the English oligopoly and their hired murderer.' While some have taken his words literally, it seems clear that he was merely reiterating an earlier observation made to Governor Lowe himself (the 'hired murderer' of the passage in question), in which it was clearly the agency of St Helena itself that was implicated: 'There is courage in putting a man to death, but it is an act of cowardice to let him languish, and to poison him in so horrid an island and in so detestable a climate.'[27] The authorities' refusal to accede to his demands to be moved from the uncongenial Longwood Plain was here claimed to be hastening his death, but in time he concluded (correctly, as it turned out at the autopsy) that he had developed the same stomach cancer that had killed his father. He became increasingly ill and on 4 May 1821 he lost consciousness. The following day, a signal was hoisted indicating that 'General Bonaparte is in imminent danger', and by nightfall, surrounded by his companions, he was dead.

Laid out in his favoured green uniform of a colonel of the Chasseurs de la Garde, Napoleon's corpse was ceremonially arranged on the camp bed he had favoured since his campaign at Austerlitz and draped in the embroidered cloak he had worn at the Battle of Marengo, so that last respects could be paid. The island population – civil, military and marine – filed past for two days. At that point (by which time early signs of decomposition had begun to set in) it was decided that before burial a mould should be taken of the face with a view to producing a death-mask. Although almost every detail of the proceedings is contested, it appears that (after some difficulties due to the lack of plaster of Paris) a piece-mould was taken by Francis Burton, a surgeon attached to the 66th Regiment who had earlier participated in the autopsy. What happened then is obscure: some say that the mould was damaged and was partially restored on the spot by Francesco Antommarchi,[28] while others maintain that the element of the mould covering the face was appropriated by the Countess Bertrand, who later employed Antommarchi to produce casts from it. Numbers of such casts in plaster and bronze, varying widely in the distance of their relationship to the original mould, are now distributed in collections around the world.[29]

A sketch of the deathbed scene (Figure 23) was also compiled by Commander Frederick Marryat, later to achieve fame as the author of maritime adventures but at the time in command of the brig *Beaver*, a recent addition to the squadron of patrol ships that had circled the island ever since Napoleon's arrival. Marryat's

[27] Forsyth 1853, I, 160.
[28] Antommarchi was a Corsican pathologist and anatomist who had been sent to the island (along with two priests) in 1820 to attend to her son's needs by Napoleon's mother. The general would have nothing to do with him (or the priests), so that it was only with Napoleon's death that Antommarchi found an appropriate role. He took part in the autopsy, but was not a signatory to the death certificate.
[29] For 'An Inventory of the Principal Death Masks of Napoleon in Public and Private Collections' by Chantal Prévot, see https://www.napoleon.org/en/history-of-the-two-empires/articles/inventory-principal-plaster-death-masks-napoleon-public-private-collections/ where primacy is given to the so-called Bertrand mask, now at Malmaison.

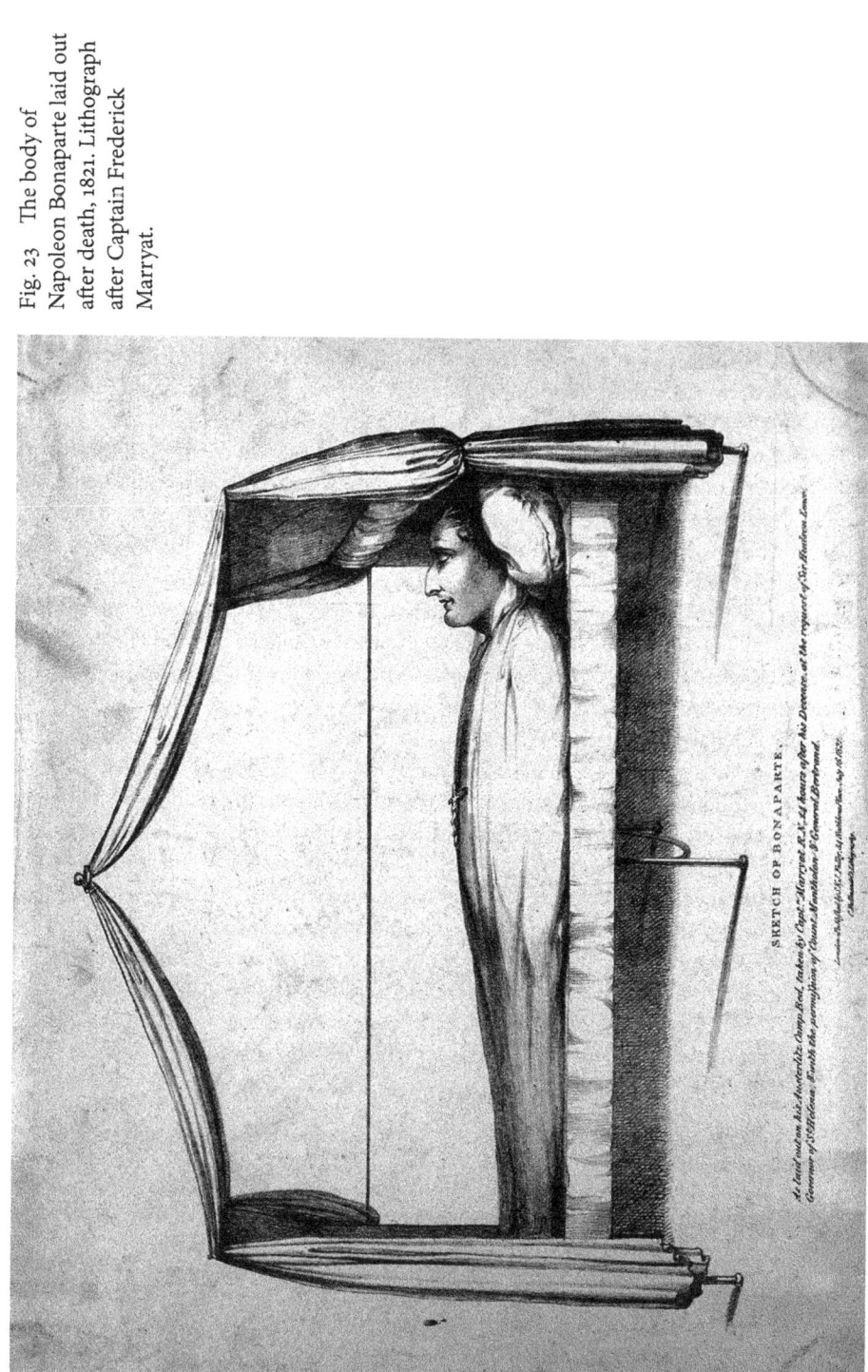

Fig. 23 The body of Napoleon Bonaparte laid out after death, 1821. Lithograph after Captain Frederick Marryat.

drawing would accompany the dispatches containing news of the former emperor's demise that left the island on the sloop *Heron* the day after its completion, for England; there all would be conveyed at speed in a coach-and-four from Plymouth to London, where the drawing was engraved and widely circulated.[30] After falling conveniently ill, Marryat had his command transferred to the sloop *Rosario*, which sailed shortly afterwards for England, where he was able to circulate an image of the funeral (Plate. 24a), which had taken place after the *Heron* had sailed.[31]

Napoleon's will had included a request that he be laid to rest 'by the banks of the Seine, surrounded by the French people, whom I love so dearly'. Not only was there no chance that such a wish would be granted by the governing regime in St Helena or indeed in Westminster, but the restored Bourbon monarchy would have been no more anxious to establish a potential focus for anti-royalist sentiment on French soil. So it was that the body was buried on the island. No fewer than four coffins, one inside the other, enclosed the corpse when it was carried from Longwood House: the first was of tin, lined with satin and padded with cotton, the second of wood, the third of lead and the outermost of mahogany covered with crimson velvet.[32] Carried from the house by twelve grenadiers of the 20th Regiment, the massive coffin was loaded onto a waiting horse-drawn hearse. As it trundled towards the chosen burial site by a freshwater spring in Geranium Valley, a guard of honour from the 66th and 20th Regiments, the Royal Artillery, Royal Marines, Dragoons and the St Helena Regiment and Volunteers – some 2,000 troops in total – lined the entire route; observers at the time commented on the piquancy of the battle-honours woven into their regimental colours – 'Talavera', 'Albuera', 'Vitoria', 'The Pyrenees' – each marking a significant reverse for the French in the Peninsular War.[33] On the final narrow path (Plate 24b), further grenadiers and naval ratings took turns to manhandle the coffin to the burial site. Salutes were fired from the harbour, answered by the shore batteries, while at the grave itself repeated salvoes were fired over the grave – three fifteen-gun salutes were fired by the artillery, followed by three volleys of musket shot.

The willow-shaded cast iron railing that modestly demarcates the burial spot betrays nothing of the massive engineering project carried out underground to receive the emperor's mortal remains – and, more specifically, to ensure that they remained beyond the reach of those who might have made relics of them. Governor Lowe himself would certainly have had a close interest in the construction (Figure 24),

[30] The French public was treated to an altogether more heroic but totally imaginary reconstruction of the emperor's end: see, for example, Horace Vernet's canvas in the Wallace Collection (inv. P575), where *The Emperor's Tomb* is located on a storm-tossed promontory, dashed with the wreckage of a vessel inscribed with the names of some of his most important battles.

[31] Warner (1953, 58) observes that, rather surprisingly, in his literary output Captain Marryat (as he later became) made no reference to his encounter with St Helena's most illustrious resident.

[32] Andrew Darling, undertaker, quoted in Levy 1998, 57. See also Fox 2021 for further documentation (and speculation) on details of the coffins.

[33] Roberts 2015, 801.

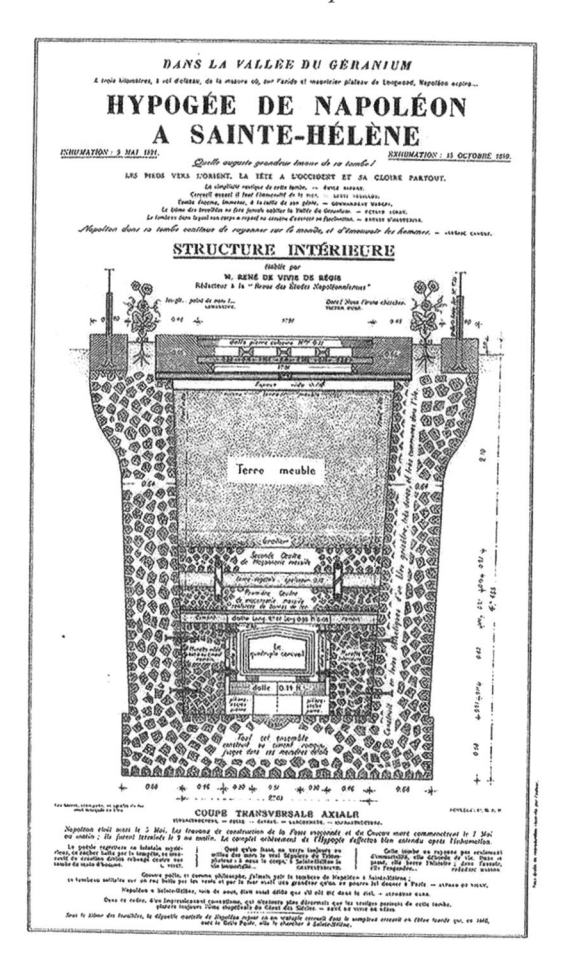

Fig. 24 Cross-section of constructional details of Napoleon's tomb.

which he described in detail a few days after the funeral in a dispatch to the Minister for War and the Colonies in London:[34]

> A large pit was sunk of a sufficient width all round, to admit of a wall two feet thick of solid masonry, being constructed on each side; thus, forming an exact oblong, the hollow space within which was precisely twelve feet deep – near eight long and five wide. A bed of masonry was at the bottom. Upon this foundation, supported by 8 square stones each a foot in height, there was laid a slab of white stone, five inches thick; four other slabs of the same thickness closed the sides and ends, which being joined at the angles by Roman cement, formed a species of stone grave or sarcophagus. This was just of depth sufficient to admit the coffin being placed within it. Another large slab of white stone which was supported

[34] Lowe to Lord Bathurst, 14 May 1821, British Library, Mss Add 20133 fol. 200 r.v.

on one side by two pullies, was let down upon the grave, after the coffin had been put into it, and every interstice afterwards filled with stone and Roman cement. Above the slab of white stone which formed the cover of the grave, two layers of masonry, strongly cemented, and even cramped together, were built in, so as to unite with the two-foot wall which supported the earth on each side, and the vacant space between this last work of masonry and the surface of the ground, being about eight feet in depth, was afterwards filled up with earth. The whole was then covered in a little above the level of the ground, with another bed of flat stones, whose external surface extending to the brink of the two-foot wall on each side of the grave, covers a space of twelve feet long and nine feet wide.

Combined with the multiple coffins, the whole ensemble exudes an air of almost pharaonic inviolability.[35]

The headstone planned to mark the grave was a final casualty of the animus that had developed between the governor and the French, with Lowe refusing to countenance use of the single 'Napoleon' due to its imperial overtones and the French refusing to accept the more prosaic alternative which he proposed, 'Napoleon Bonaparte'. As a result, the stone ever remained blank.

Although the spot chosen for the burial appeared at the time to be suitably tranquil and withdrawn, it suffered with the passage of time and from the large number of visitors attracted to what quickly became the island's principal tourist attraction. One such sightseer drawn to the tomb was Charles Darwin, who came to pay his respects in 1836; he found it now 'situated close by cottages & a frequented road [so that it] does not create feelings in unison with the imagined resting place of so great a spirit'.[36]

All the former residents at Longwood had long since been repatriated by Darwin's day, but Lord Liverpool's conviction that Napoleon, 'being so far from the European world, would soon be forgotten' would prove seriously ill-founded. With the advent of more sympathetic regimes in both England and France, the prospect of honouring the emperor's last wishes finally came to fruition in 1840, when a formal request from the French for the return of Napoleon's remains was agreed and an embassy led by the son of King Louis-Philippe sailed to St Helena to complete the formalities.[37]

---

[35] Lowe's own monument, in St Mark's Church, Mayfair, alludes to the thankless task he carried out on the island and the long-awaited redemption of his reputation: 'He was selected for the onerous post of Governor of St Helena, during the captivity of Napoleon. His obedience to the orders of Government in the fulfilment of this harassing duty, earned him the approbation of his Sovereign, but exposed him ever afterwards to persecution and calumny, and more than once caused his life to be endangered. History will do justice to a brave and zealous officer, a true and generous friend and an upright and faithful servant of his country.' A passionate vindication of every aspect of Lowe's character and conduct, published almost a century after the events in question, characterizes him as 'the real martyr of St Helena' (Pillans 1913).

[36] Darwin 1933, 410.

[37] Antipathy to recognition of the title of emperor had long since evaporated amongst British officialdom, so that Napoleon left the island in full possession of the honours

The body was laboriously exhumed from its masonry vault (the site of which was by then less visited and showing signs of neglect) and was transferred to an even more impressive array of six caskets for ceremonial return to Paris where, on 2 December 1840 – the anniversary of his coronation – Napoleon received a second funeral more fitting to his imperial status: a million spectators are judged to have lined the route as the coffin was drawn through the streets to be enshrined in the gigantic porphyry sarcophagus at the Hôtel des Invalides in which it rests today.

As for Longwood, the passage of time quickly obliterated any hint of its former prestige; within a decade it had reverted to an agricultural storehouse and stables. Continuing his search in the general's footsteps, Darwin observed: 'With respect to the house in which Napoleon died, its state is scandalous, to see the filthy & deserted rooms, scored with the names of visitors, to my mind was like beholding some ancient ruin wantonly disfigured.'[38]

A change of fortune finally came in 1858 when Longwood, along with the pavilion at The Briars and the site of Napoleon's tomb, were vested in His Majesty Napoleon III and his heirs. Today there remain two utterly irreconcilable views of the legality and morality of his detention on the island. All the decision-making as to its implementation was undertaken almost 5,000 miles away in the corridors of Whitehall, but for those who know of the island in no other context St Helena will always be the place of Napoleon's detention and ultimately his death. The popular image of the erstwhile emperor staring belligerently out to sea from the cliffs – although based on no contemporary original[39] – seems destined to remain the most iconic image from this episode in the island's eventful history.

---

that had been conspicuously denied him at his arrival. In 1858 Longwood and its 33-acre estate were sold to the French, along with the site of the tomb. The garden pavilion at The Briars was bequeathed to France in 1859 by a great-granddaughter of William Balcombe and today the three sites remain French territory, forming the Domaines Nationaux à Sainte-Hélène.

[38] Darwin 1933, 410.
[39] The best-known image is that by Paul Delaroche; his work survives only as an oil sketch in the Royal Collection (RCIN 404876), begun over thirty years after Napoleon's death but made widely known through an engraving by C. W. Sharpe, c.1860.

8

Later detainees, 1800s and 1900s

The death of Napoleon and the consequent withdrawal of perhaps half of the island's population – all those who had been involved more or less directly in his incarceration – brought about one of the recessions in the island's economy that followed repeatedly in the later years of the nineteenth century. When Darwin passed through the island in 1836 the guide he employed probably spoke for the majority of the population when he looked back on the 'fine times' of the Napoleonic interlude as something of a golden age.[1]

For the British government the role of the island in isolating the former emperor from contact with the outside world was judged to have been – with the possible exception of the cost of it all – entirely satisfactory, so it was perhaps inevitable that the name of St Helena should be invoked again whenever the need arose for secure and suitably remote accommodation.[2] For nineteenth-century politicians and colonial administrators such a choice would have been considered expedient and routine. As late as 1968, when King Goodwill Zwelithini acceded to the throne of the Zulu nation at the age of fifteen and was judged to be at risk from his opponents until he reached his majority, it seemed appropriate that until the moment for his coronation in 1971 a refuge could be found for him on St Helena.[3] His grandfather, Dinuzulu, had been accommodated there more formally some eighty years earlier (see below). More extraordinary is the survival even today of a perception in Whitehall that the island might still reasonably function as an appropriate sanctuary – or a prison – in which politically problematical individuals might conveniently be sequestered.

---

[1] Darwin 1933, 411. The downturn in the island's prosperity had been exacerbated by its transfer to government control in 1834 and the consequent loss of so many EIC benefits and emoluments.

[2] During the earlier years of the Napoleonic wars the island had already accommodated some 300 prisoners of various nationalities, said to have been taken out of Dutch ships in particular (Gosse 1990, 225).

[3] Obituary, *The Times*, 13 March 2021.

Later detainees, 1800s and 1900s

Policing the slave trade

Having witnessed the introduction of slavery from the earliest years of its human occupation and having itself become a regular consumer of enslaved Africans and others – it will be remembered that the EIC had long accepted the principle that no plantation in the tropics could be sustained without them – there was a certain piquancy in the choice of the island as a key base in enforcing its prohibition in later years. Within a year of the passage of the Act for the Abolition of the Slave Trade (1807),[4] the Royal Navy had established a West Africa Squadron with the aim of suppressing the trade by sea. From an initial two small sailing ships with limited authority against foreign shipping, it gradually expanded by the mid-nineteenth century to twenty-five vessels (including a number of paddle steamers), crewed by some 2,000 British seamen and 1,000 'Kroomen' – experienced African sailors. It came to form an effective unit, renamed the Preventative Squadron, and was estimated to have captured some 1,600 slave ships and to have freed perhaps 150,000 Africans.[5]

From the early 1840s onwards the West Africa Squadron cruised near the coast and mounted longer-range patrols in the South Atlantic in search of ships illicitly attempting to continue the supply of slaves to the Americas. Within a decade some twenty-four cruisers had made St Helena their base.[6] A monumental column topped by a draped urn, standing in the Castle Gardens in Jamestown, bears witness to the exigencies suffered by members of the crew of just one of these vessels, HMS *Waterwitch*, a 10-gun brig – the first vessel of the fleet to bring a captured slave ship to the island and which accounted for a record number of forty-three others.[7] Of the seamen memorialized here, 'The greater number died while absent in captured slave vessels. Their remains were either left in different parts of Africa or given to the sea.' The inscription tells us that, 'This island is selected for the record because three lie buried here and because the deceased as well as their surviving comrades ever met with the warmest welcome from its inhabitants.'

A Vice-Admiralty court was established on St Helena for the prosecution of those apprehended in the illegal trade:[8] several hundred ships seized in the process,[9] mostly

---

[4] The Act of 1807 had outlawed the trading of slaves within the British Empire, although ownership of slaves persisted until enactment of the Slavery Abolition Act (1833).
[5] Many of the latter chose to settle in Sierra Leone (the first British colony in West Africa), in order to avoid being re-enslaved at home.
[6] Van Niekerk 2009, 2.
[7] *Waterwitch* had been built on the Isle of Wight as a racing yacht for George Hamilton Chichester, Lord Belfast, before being taken up by the Royal Navy and converted for her new role. See Pearson 2016.
[8] The governor of the day, General George Middlemore, was appointed Vice-Admiral for the purposes of this court, whose officers, under Judge Charles Hodson, were drawn from the resident population. For a detailed account of the constitution of the court and of the cases brought before it, see van Niekerk 2009. See also Van de Velde 2011.
[9] Van de Velde (2011, 12) gives a total of 425 ships tried at the Vice-Admiralty court between 1840 and 1867, which between them had transported between 21,500 and 25,000 enslaved men, women and children.

bound in earlier years for Brazil and later for Cuba,[10] were broken up on the island on the orders of the court, their construction materials amounting to some tons of timber and iron intensively recycled by the resource-starved islanders and adding significantly to the economy. A few of the better-quality captured vessels – notably the fast 'Baltimore clippers' that could outrun the British vessels – were taken up into service.

Treatment of the unfortunate human cargoes rescued from this inhumane trade was a problem not easily resolved.[11] J. C. Mellis describes a visit he made to one of these vessels, brought under escort into Rupert's Bay:

> A visit to a full-freighted slaveship arriving at St Helena is not easily to be forgotten; a scene so intensified in all that is horrible almost defies description. The vessel, scarcely a hundred tons burthen at most, contains perhaps little short of a thousand souls, which have been closely packed, for many weeks together, in the hottest and most polluted of atmospheres ... One's sensations of horror were certainly lessened by the impossibility of realizing that the miserable, helpless objects being picked up from the deck and handed over the ship's side, one by one, living, dying, and dead alike, were really human beings ... Many died as they passed from the ship to the boat, and, indeed, the work of unloading had to be proceeded with so quickly that there was no time to separate the dead from the living.[12]

As an initial step, 'Liberated African Depots' were established at Lemon Valley (short-lived, and closed in 1843)[13] and at Rupert's Valley (Figure 25); the latter operated until 1867 in an uncomfortable cycle of desperate overcrowding as captured slavers – some of them with up to 800 on board – arrived unannounced, and periodic abandonment and dilapidation when no seizures were made. The human cargoes of these vessels, very often in a skeletal state from the inhuman conditions they had endured, were disembarked, their multiple ailments treated where possible, the more fortunate resettled and the unrecoverable buried on the island. In total some 26,000 'recaptives' were brought to the island over the years, for many of whom there would be no recuperation, as indicated by the testimony of Bishop Robert Gray, who visited the Rupert's Valley encampment in 1850:

---

10   Van de Velde (2011, 7), quoting data gathered by W. G. Tathams in the Governmental Archives at Jamestown in the 1960s, reports that some 80 per cent of the vessels brought to St Helena had been seized along the coast of Angola rather than on the high seas.
11   For an account of the continuing miseries suffered by recaptives on board the seized vessels, and of the varying degrees of success with which crews managed to inject some humanity into their assignment, see Wills 2019.
12   Mellis 1875, 30–31.
13   The Lemon Valley camp was never developed to the extent of that in Rupert's Valley. Shortage of accommodation here made it necessary for new arrivals to be held in quarantine in hulks anchored off the coast, adding to the misery of their experience.

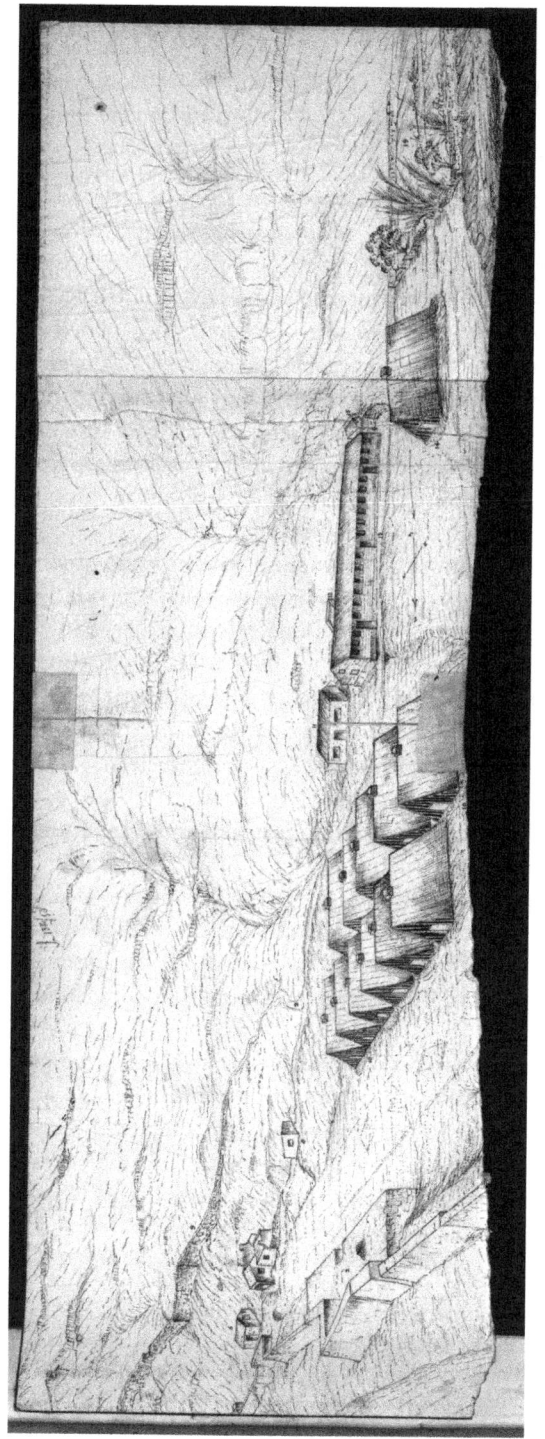

Fig. 25 Drawing (fragmentary) of the Liberated Africans' camp, Rupert's Valley, c.1858, bound into the Colonial Office's correspondence volume for that year.

> If anything were needed to fill the soul with burning indignation against that master-work of Satan, the Slave-trade, it would be a visit to this institution. There were not less than 600 poor souls in it ... of these more than 300 were in hospital; some affected with dreadful opthalmia; others with severe rheumatism, others with dysentery, the number of deaths in the week being twenty-one.[14]

About one-third of the number 'rescued' are thought not to have survived: in addition to the afflictions noted by the bishop, yellow fever, smallpox, measles, beri-beri and onchocerciasis ('river blindness') were rife amongst each new wave of arrivals, testing the capacities of the resident surgeon and his overworked staff to the limits. To the inadequacies of contemporary medical treatments were added the difficulties of persuading the African population to submit to many of the practices prescribed by the surgeon. Even the uncertainties of the island's food supplies may have contributed to the process of attrition (although in this respect their lot is said to have been no worse than that of the garrison). The Rupert's Valley cemetery alone, partially excavated in 2007–08 ahead of construction of an access road for the airport, is estimated to have received some 8,000 bodies (Figure 26).[15]

The report on those excavations presents the findings made on the ground, together with a digest of the documentary records relating to the camp. The principal accommodation provided was in the form of eleven 'wooden tents' arranged in two rows – simple structures resembling large ridge-tents, their roofs sloping to the ground, with an isolated twelfth tent housing recuperating patients from the more substantial timber hospital building. The administrative offices were located in the stone-built structures of the military lines defending the foot of the valley.

For the 'liberated Africans' who survived their introduction to St Helena, an attitude seems to have prevailed in the administration that they could legitimately be set to work on various schemes within the island while awaiting resettlement – bringing an improved water supply to Rupert's Valley, repairing the roads or clearing invasive blackberries and gorse from agricultural land. Like the cyclical nature of the occupancy of the camp, relations between the Africans and the European administrators varied from time to time – from engagement in sporting competitions to the threat of mutiny.

Internment on the island seldom proved a prelude to return to a place of origin. DNA analysis of the remains of twenty individuals from the cemetery (where some 325 interments were uncovered in the excavation) has revealed that they originated largely in northern Angola and central Gabon (though the current lack of comparative data from Africa limits the precision of the findings).[16] Some

---

14   Reproduced in Gosse 1990, 320.
15   Pearson *et al.* 2011. On 7 September 2022, *The Times* reported that the remains of 325 individuals had been ceremonially reinterred on the site, each in a wooden casket made by school children on the island. Plans are in hand for the construction of an accompanying memorial and interpretation centre.
16   Pearson *et al.* 2011; Callaway 2016.

Fig. 26 Excavations taking place at the cemetery associated with the Liberated Africans' camp at Rupert's Valley, 2007–08.

resettlement of rescued Africans took place to Sierra Leone and the Cape, but as many as 10,000 are thought to have been compulsorily carried on to the British West Indies or to British Guiana – no longer 'recaptives' in the terminology of the time and now destined for nominal freedom, but in reality for a life of continuing hardship as indentured labourers in the sugar trade and other industries.[17] Others – perhaps up to 1,000 – remained on St Helena where they mostly found work as servants in the islanders' households, while male children were apprenticed into various trades and ultimately became absorbed into the fabric of the population.

The ships policing the slave trade were finally withdrawn in 1864, and the Liberated African Depot was closed a decade later. It is a measure of the precariousness of the island's economy that even such a pitiable development had brought a degree of temporary prosperity to those islanders who supplied the needs of the camps; with their closure, another relapse set in.

Zulu chieftains in the mid-Atlantic

With the passage of a further twenty years or so, the British government again found itself in need of a detention centre – not exactly a prison, since the island's very remoteness could be relied upon to provide a large measure of security. The occasion was prompted by an uprising in 1888 in Zululand, which had been annexed by the British the previous year, following the Anglo-Zulu War of the previous decade. In the post-war settlement the British found themselves opposed by an army under the paramount Dinuzulu, son of Cetshwayo, the last king of the Zulu nation; the army was routed and a hunt for Dinuzulu followed, led by Captain Robert Baden-Powell. Although Dinuzulu managed to escape across the border into the Transvaal, he returned a year later and gave himself up to the British authorities, who promptly sentenced him to ten years' imprisonment for treason, to be served on St Helena.

Like Napoleon, Dinuzulu did not arrive alone: thirteen persons disembarked at Jamestown on 25 February 1890, including two of his uncles, named Ndabuku and Tshingana, together with their respective wives, two interpreters and four male and two female attendants. All were accommodated for a time at Rosemary Hall in St Paul's and later in upper Jamestown, before Dinuzulu and his entourage moved to Francis Plain House in St Paul's. He wrote of the family's experience there:

> We live in a very large and very nice house. It is cool and away from the mass of people. The house is situated on a hill and we live alone. We see the Governor of St Helena very often. He is very nice and visits us frequently. This is a very large place. We were wrongly informed when we were told we were going to live on a small rock.[18]

---

[17] Gosse 1990, 310.
[18] See http://sainthelenaisland.info/exiles.htm, to which much of this section is indebted.

Fig. 27 Dinuzulu takes his leave of the Governor of St Helena, December 1897.

The separation of Dinuzulu from his uncles was not a punitive measure; in fact, the regime under which all of them were detained seems to have been relaxed in the extreme, with no limits placed on their ability to go where they pleased or to engage with the resident population. Rather, it reflected a growing distance that developed between the various detainees in their attitudes to captivity. While the uncles proved resistant to all attempts at acculturation – they refused to make use of chairs, tables or beds, and shunned European clothing except when they left their house (one of the few conditions of their being allowed out) – Dinuzulu proved receptive to all that the island had to offer him. He and his growing family (he and his wives had eight children on the island, two of whom died there) developed a fine appreciation of European dress and culture, and they mixed easily in society (Figure 27): he was taught to read and write and to play the piano; he was on friendly terms with the resident bishop and clergy, and was indeed baptized into the Christian faith. His captors evidently approved of the transformation: Dinuzulu was judged to have become 'more refined, his gentlemanly manners and bearing promising well for the tribe over which he may hold sway'.

By the time the party left the island on Christmas Eve 1897 it had swollen in number to twenty-five. St Helena said goodbye to them with regret, not least for the loss to the local community of £1,000 a year that had enriched them from the presence of the free-spending Dinuzulu.

On his return to his people Dinuzulu was duly restored to a position of power by the British, but their high hopes for him were dashed when another rebellion earned him a further four years' imprisonment in his homeland in 1908. The sentence was ended after two years when his friend General Louis Botha (who had always believed that Dinuzulu's detention was unjust) became prime minister of the Union of South Africa and restored his liberty.

Boer War prisoners

A more significant boost to the island's economy came within a few years when much larger contingents of prisoners of war from the Boer War campaigns began to arrive on the island, to be installed in a camp on the windswept Deadwood Plain that had so displeased Napoleon and his entourage.[19]

The war in southern Africa had pitched the British against the two independent republics of Transvaal and Orange Free State. Hostilities were declared on 11 October 1899 and continued for the following two and a half years. In the course of the conflict the British forces found themselves with so many prisoners on their hands (a total estimated as high as 25,000) that in order to isolate them securely (and to avoid the logistical problems of having so many additional mouths to feed within the disputed zone), a decision was taken that they should be sent to prisoner-of-war camps in British possessions overseas – in India, Ceylon, Bermuda and, most notably, in St Helena.

The first contingent arrived at Jamestown on 10 April 1900, when the troopship *Milwaukee* arrived with 514 prisoners on board; their disembarkation was delayed for a week when the ship was placed in quarantine due to an outbreak of measles. Although the forbidding appearance of the island on the approaches must have been intimidating for the prisoners, the regime they encountered seems from the moment of their arrival to have been notably mild in character. A notice had recently appeared in the pages of the *St Helena Guardian*:[20]

> His Excellency the Governor expresses the hope that the Inhabitants will treat the prisoners with that courtesy and consideration which should be extended

---

[19] Susan O'Bey, 'Saint-Helena – prison island', at https://www.napoleon.org/en/history-of-the-two-empires/articles/saint-helena-prison-island/

[20] *St Helena Guardian*, 12 April 1900.

to all men who have fought bravely in what they have considered the cause of their Country and will help in repressing any unseemly demonstration which individuals might exhibit.

By all accounts the prisoners presented a sorry sight as they disembarked. The same newspaper judged them

> a motley crowd of beings of all ages, from boys of 14 to men of 60, some clean and decently clad, others poorly clad, dirty and unkempt, and sickly-looking, each with a dirty haversack, water kettle or bottle, or string of drinking pots and pans, some with bundles of clothing wrapped in blankets.

While the local populace lined the streets (observing punctiliously the governor's strictures as to good behaviour) the prisoners began to move towards their camp under military escort (Figure 28). The three-hour march up the steep, uneven road to Deadwood proved too much for many of the men, even with repeated rest periods; numbers of the most sickly and exhausted had to be carried by ambulance. Their entry to the compound that was to be their home for the next two years was enlivened by the fife and drum band of the West India Regiment.[21]

The prisoners were led inside a fence of barbed wire several hundred metres square containing a good number of canvas tents in which they were to be housed for the duration of their stay on the island. The Boers were reported to be in good spirits; laughing, joking and playing games of all sorts. The following day they had a meeting among themselves where they elected one of their group to be a sort of Captain, to whom they could report grievances, and who would represent them. A further batch of nearly 400 prisoners arrived a fortnight later on 26 April, and within weeks their numbers had risen to over 4,500,[22] to be joined by over 1,000 more at the beginning of 1902.

Only the senior officer in the initial contingent was allowed a measure of deference. General Piet Cronjé and his wife and their servants disembarked on 13 April and were driven to be formally received by Governor Sterndale before being carried on to Kent Cottage in Half Tree Hollow, where they were to stay for the remainder of their time on the island. Philip Gosse tells of General Cronjé's dissatisfaction when he found that his appointed cottage (near High Knoll fort) was to be guarded by foot soldiers (with the guard changed every day) rather than the mounted guard he considered appropriate to his rank. To oblige him, a few members of the St Helena Volunteers were

---

[21] Interestingly, when an earlier manifestation of the St Helena Regiment was being run down in the 1860s, a question in the House of Lords as to its future had drawn the response that 'it was intended to be absorbed by the gradual draught of men into other regiments' and that 'a considerable number of the officers had been already transferred to a West India regiment' (*Hansard*, vol. 171, 22 June 1863).

[22] The number included the grandson of Paul Kruger, president of the Transvaal.

Fig. 28 Arrival of a contingent of Boer prisoners of war in Jamestown, 1900–01.

quickly schooled in equestrianism so that they could accompany the general about the island, but had to suffer the indignity of having their shortcomings exposed by the hard-riding Cronjé before the assembled ranks of his South African horsemen.[23]

A second Boer general, Ben Viljoen, arrived in February 1902 on board the *Britannic*. Evidently he was treated with less ceremony and was initially confined in the camp; after protesting to the commandant, Major S. H. Marden, he was allowed to stay in a small house, Rose Cottage, outside the perimeter fence.

As well as the Boers themselves – for the most part mild-mannered farmers rather than professional soldiers, many of them shown in contemporary photographs wearing sober suits and straw hats – the prisoners included numbers of foreign volunteer fighters of a dozen nationalities, amongst whom the Hollanders were seen as particularly disruptive. Men identified as particular trouble-makers were singled out for consignment to High Knoll fort, which had long been used as a military detention centre.

For the most part the prisoners were housed in canvas tents (Plate 25), though with time the more enterprising constructed huts and shelters for themselves, many incorporating old paraffin cans and biscuit tins, flattened out and fastened together to keep the weather at bay. Canteens were organized by the prisoners themselves. They also formed clubs, musical groups and teams for various sports; there was even a camp newspaper, *Kamp Kruimels* (later *Die Krijgsgevangene*). Correspondence with the outside world was allowed, but was subject to censorship.[24] (It was interrupted only when an outbreak of bubonic plague in South Africa in 1901 resulted in the quarantining of all incoming ships.) The more industrious prisoners turned their hands to whittling and carving (Plate 26), producing tobacco pipes, walking sticks and models carved from timber (recycled or dead wood – particularly ebony, olive and redwood – gathered in the countryside). Victoria Heunis has uncovered an astonishing photograph of H. T. Siglé's woodcarving establishment at Deadwood (Figure 29), complete with a small windmill that powered a wood-turning lathe within – signifying a remarkable degree of latitude towards this minor industry within the military encampment – but most of the production was undoubtedly achieved with little more than knives, chisels and boring implements.[25] Bone

---

[23] A low point occurred on one occasion when the newly recruited cavalry escort had to be helped to remount their horses by the prisoners, who held their rifles while they regained the saddle (Royle 1998, 57).

[24] For a list of the 'Rules Regulating the Correspondence of Prisoners of War' and the low character of the chief censor at Deadwood, F. W. Alexander, see Royle 1997.

[25] Heunis 2019, 67–68, fig. IV, 5. Royle (1998, 64) documents a visit to the camp by the reformer Alice Stopford Green, who was so eager to forward the prisoners' interests as to lobby for the regular supply of raw materials and to set herself up as their agent, arranging the profitable export of numerous walking sticks, pen holders, cigar holders and egg cups.

Fig. 29 The well-established workshop of H. T. Siglé in the Deadwood Boer prisoner-of-war camp, 1901–02. The windmill is said to have driven his turning lathe.

provided a further popular raw material, perhaps mostly from kitchen waste but also whale bone, no doubt supplied by the prisoners who worked in the whaling industry's shore-based processing plant on the island (see Figure 10); the corpses of whales washed up on the shore periodically provided a further source of bone and teeth for craft production – snuff-boxes, walking sticks, brooches and small ornaments being particularly favoured.[26] Heunis mentions a further source of bone that brings this particular industry into a decidedly macabre light. A certain amount of scavenging was carried out in the area of the former 'Liberated Africans' camp and its associated cemetery at Deadwood, producing the raw material for brooches and other ornaments made of human bone.[27] Lesser crafts relied on metal, stone and shell. When an exhibition of the island's industrial arts was organized at the Castle in 1900 the prisoners contributed so numerously that an additional room had to be provided for their handiwork (Figure 30). A dozen individual producers are named and provided with brief biographies by Heunis.[28]

[26] Heunis 2019, 81–82.
[27] Ibid.
[28] Ibid., 107–17. A report in *Die Krijgsgevangene*, 8 June 1901, indicates that some part of this production went towards the establishment of a Widows' and Orphans' Fund for the families of the prisoners. A committee was formed 'to collect all

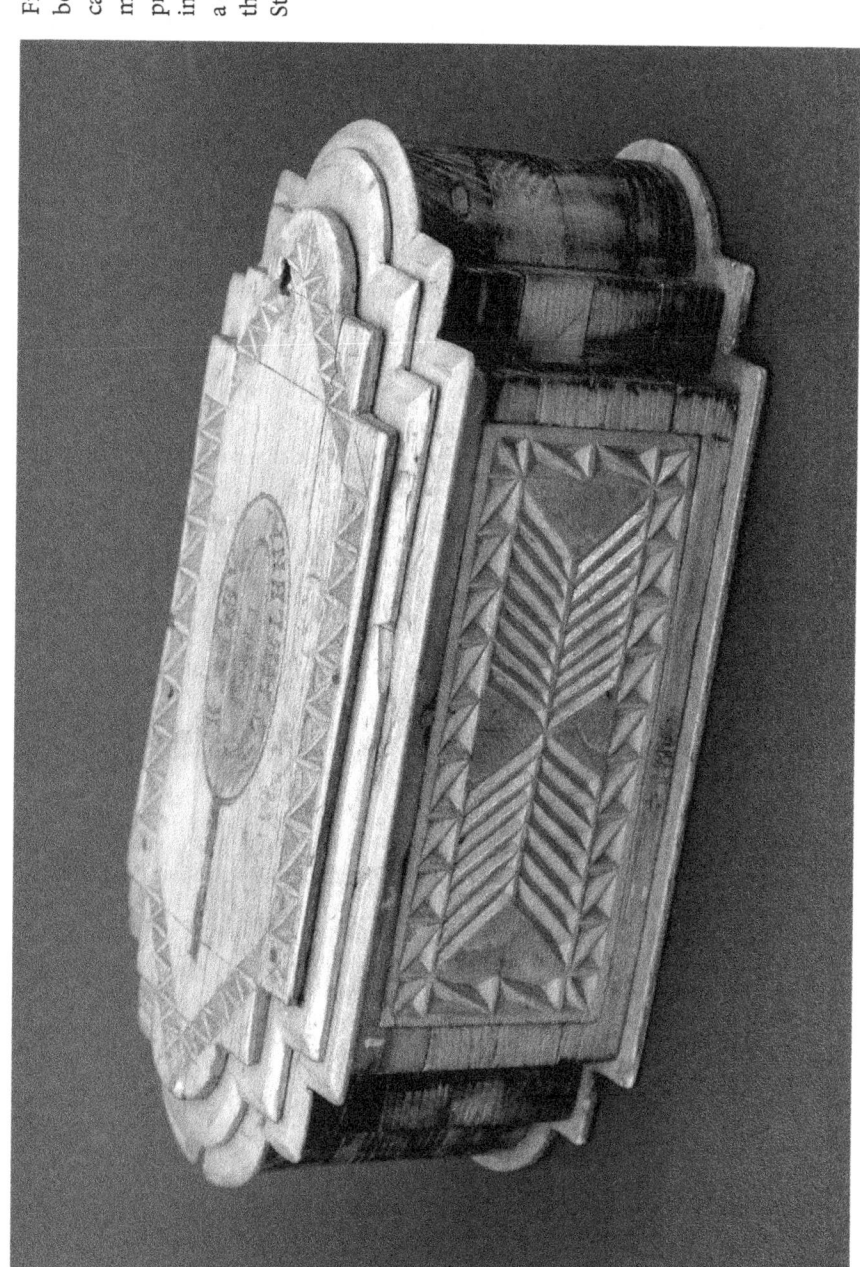

Fig. 30 Money box with chip-carved ornament made by a Boer prisoner of war, inscribed within a cartouche on the lid, 'K. Gey, St Helena. 1901'.

Barbed wire fences marked the extent of the Deadwood camp, patrolled outside the perimeter by soldiers. Parades were minimal, beyond a roll-call at the end of each day. Relations between captors and prisoners could at times be comparatively amicable, although some of the guards were undoubtedly harsh and brutal; one of the few occasions when the situation degenerated seriously resulted in the death of one inmate, shot by a sentry for allegedly throwing stones at him. The garrison was made up of some 1,500 British troops, mainly from the 4th (Militia) Battalion of the Gloucestershire Regiment.[29] Indeed the respective lots of the guards and the inmates probably did not differ greatly from each other: in seeking to identify possible causes for an outbreak of 'beri-beri' at the camp, a report in the *British Medical Journal* observes that the rations of the troops and the prisoners were identical – 'fresh meat four days a week, tinned meat two days, vegetables and potatoes on alternate days, bread, sugar, and coffee' – while conceding that 'it is true the vegetables were compressed, and the potatoes were often not too good'.[30] Needless to say, the soldiers were subject to the same 'iron-handed' discipline exercised by the camp's commanding hierarchy.[31]

The inmates themselves, by contrast, became increasingly factious, leading to hostilities so marked that further camps were established in order to separate the most entrenched groups. In 1901 a camp was established some five miles away at Broad Bottom to accommodate natives of Orange Free State, who had developed a mutual antagonism towards those from Transvaal.[32] At Deadwood itself, a second camp (designated the 'Peace Camp') was set up for the growing numbers of prisoners who formally declared themselves in favour of peace or who had decided to seek British citizenship; these consequently had found themselves alienated by the other

---

kinds of curios and articles made by the prisoners of war, to pack and send them to Holland where they will be sold ... With the last mail ship the first consignment is already on its way, containing 550 pen holders, 50 brooches, a number of sliding, secret, spectacle, match and other boxes and many other wood and bone articles too many to describe.' A year later the same journal (15 June 1902) recorded that 'our camp's business and faculties have increased and with exercise and gradual accumulation of better tools, articles have been made of much better quality and shape as when the above mentioned exhibition was held' (http://sainthelenaisland.info/boerprisonersnationaltrust.pdf).

[29] The Gloucestershire Regiment had indeed received the surrender of General Cronjé and 4,000 of his men following the Battle of Paardeburg.

[30] *British Medical Journal* 2 no. 2181 (1902), 1258. There were some who blamed the introduction of the disease on the Boers, but Shine (1970, 22) asserts that it was already endemic at a low level on the island – camp life merely gave it an ideal breeding ground. As for the quality of the military diet, Royle (1998, 61) records that troops of the West Indian Regiment mutinied in 1900 in protest at the meagre rations they received.

[31] Royle 1998, 63.

[32] The prisoners at Broad Bottom evidently were equally industrious: Heunis (2019, 196, fig. VIII, 5) reproduces a photograph of an exhibition of their products from 21 June 1902.

inmates.³³ It was proposed to open yet another camp at Francis Plain, where troops of the Gloucestershire Regiment were already encamped, but local opposition to the loss of the island's principal recreation ground led to its withdrawal. Amongst the most disaffected there were a few desultory attempts at escape from the island – some ingenious and others doomed from their inception – but all proved in vain.

Prisoners with particular skills quickly found their way into the island community where their talents could be put to good use; many others found employment as household servants or working on the roads. A number even went into business on their own account: records are found of individual prisoners becoming established as a pawnbroker and an auctioneer, while others set up a coffee house and a brewery;³⁴ a few found work with the Eastern Telegraph Company. Numbers of prisoners were allowed to stay with families who gave them employment, or in satellite camps set up in Jamestown itself. Inevitably, some became romantically involved with local girls and at least two marriages are recorded.

From the islanders' point of view, the prisoners of war and the garrison drafted in to watch over them constituted an enormously valuable market for their produce – meat, fish, milk, fruit, vegetables, etc. Susan O'Bey conjures up a picture of Jamestown as 'a hive of activity with hundreds of donkeys loaded with provisions for the camps and numerous mule wagons and bullock teams traversing the narrow streets'.³⁵ Steamships bustled in and out of the roads, while a permanent naval presence providing security patrols added further to the vibrancy. For a period the economy was more buoyant than at any time since the heyday of Napoleon's captivity.

Within the camps a greater threat than factionalism came from the diseases that inevitably overtook large concentrations of men living in less-than-ideal conditions. By May 1902 enteric fever had become established to a worrying degree; the troops were also infected, as indeed was the island population as a whole. Amongst the prisoners, thirty-one succumbed to the disease. The establishment of a separate isolation camp eventually helped to bring the outbreak under control.

Extraordinarily, the Anglican Church refused to bury the dead prisoners in its established cemeteries on the basis that they were 'enemies of the King'. Instead it fell to the Baptists (whose church had always been freely open to the Boers) to see that they received a Christian burial in a cemetery established on their own ground at Knollcombes. A total of 180 Boers died on the island, most of them of illnesses; the oldest was aged seventy-four – a further indication of the wide spectrum of everyday society swept up into the hostilities.³⁶ The individual graves

---

33 Ultimately, when General Cronjé himself took the oath of allegiance he took care to keep his guard with him, fearing the resentment of some of the prisoners: he left St Helena for the Cape on 22 August 1902 along with almost 1,000 prisoners in the transport *Tagus*.
34 Susan O'Bey at https://www.napoleon.org/en/history-of-the-two-empires/articles/saint-helena-prison-island/
35 Ibid.
36 Royle 1998, 58.

Fig. 31   Boer War cemetery at Knollcombes. Almost 200 prisoners of war are buried in the remarkably sloping graveyard, their names recorded on the granite obelisks at the bottom of the slope.

are neatly laid out on a steep slope (Figure 31); each is marked with a number, the key to which (giving the name and age of the deceased) is provided by two obelisks of reddish-brown granite, one of them erected at the time by the prisoners and the second put up by the South African government in 1913.

For the more fortunate survivors the end of their internment was signalled by the signing of the peace on 31 May 1902. Intent on securing an early return home, some 300 of them signed an Oath of Allegiance in Jamestown on 17 June, and ten days later the first 470 of them sailed for home on the SS *Canada*. The last of the prisoners left on 21 October aboard the SS *Golconda*; within three months the camps had been demolished and the material remains sold by auction.

St Helena could look back once again on a period when the island's fortunes had been boosted – this time to the tune of about £10,000 a year – by the misfortune visited on others who had never anticipated making an acquaintance with the island.

Sayyid Khalid bin Barghash Al-Busa'ldi

Having proved itself an accommodating repository for political prisoners, St Helena continued to be called upon in that role well into the twentieth century. One of the most exotic detainees was Sayyid Khalid Barghash Al-Busa'ldi (1874–1927), who had ruled in Zanzibar for a brief three days, 25–27 August 1896, after seizing power at the death of his cousin (whom many suspected to have been poisoned by Sayyid Khalid). Since 1890 Zanzibar had been a British protectorate, following the conclusion of a treaty that included a provision that no change of ruler could take place without British approval. To judge from the report of one official, it was clear that Sayyid Khalid (Figure 32) was unlikely ever to win such approval:

> Young Khaled has all the temper of his father Burghash, without his intelligence. He is of a sullen and very obstinate temperament, illiterate [in European languages], inordinately proud, and with a hearty dislike of European innovations. If he were Sultan, he would not only be extremely difficult to control, but would very likely give rise to grave scandals by reason of acts of cruelty and oppression.[37]

Little excuse was needed for the British to take action at his seizure of power: the shelling of Zanzibar by the Royal Navy brought about a surrender within forty minutes (but at the cost of the lives of an estimated 500 defenders). Sayyid Khalid himself managed to escape capture and claimed asylum in the German consulate; from there he was whisked by gunboat to neighbouring German East Africa, where he lived in some luxury and with full acknowledgement by the Germans as sultan for the next twenty years. With the defeat of the regional German forces in World War I and the renaming of the state as Tanganyika Territory, now under British control, time had run out for Sayyid Khalid: he was arrested in the Rufiji delta to the south of Dar-es-Salaam in February 1917 and four months later was exiled to St Helena.

The arrival of the former sultan and his entourage of seventeen followers (plus three political exiles from Kenya) evidently caught Governor Cordeaux at something of a disadvantage, as he communicated to the Colonial Office:

> ... in the entire absence of any papers or instructions from East Africa, beyond a nominal roll sent with the party, their arrival was entirely unexpected and I am still unaware of the degree of confinement they should be subjected to. I have thought it best, therefore, to place them in military custody under an officer's guard in disused quarters in Jamestown Barracks, allowing them free use of the enclosed parade ground adjoining, but no liberty to go beyond this ... In view of the local excitement, caused by their arrival, I took steps to ensure that they should not be referred to in the local newspaper.[38]

Few restrictions were placed on the detainees within their quarters, but contact with the civilian population remained strictly forbidden. Nonetheless, the party struggled to adjust to their new mid-Atlantic surroundings. In December 1918 Sayyid Khalid petitioned King George V on the matter of his detention: 'We cannot live on this island because we are people from East Africa, the climate of this island is not agreeable to us, and there are no Muslims.' A second petition addressed to the governor the following year elaborated on his concerns:

---

[37] Sir Gerald Portal, quoted in Frankl 2006, 162.
[38] Frankl 2006, 165.

Fig. 32   Sayyid Khalid Barghash Al-Busa'ldi, Sultan of Zanzibar (1874–1927).

Later detainees, 1800s and 1900s

> I, being a Muslim, beg to state to your Excellency that there are no Muslims or mosque or a Muslim who makes the necessary Muslim's affairs when one dies, to be found on this island of St Helena. Therefore I beg the British government ... to remove me to my native country Dar es Salaam to live on my plantation, or to any other Muslim country ... If my above request is not granted I beg the government to permit my women and children to return to their home so that they may not neglect their religious usages for my sake.[39]

These and further requests were eventually heeded and in 1921 the entire party was embarked on the SS *Cawdor Castle* on the first leg of their relocation – not to Zanzibar but to the Seychelles, at the time another favoured detention centre for political prisoners. In time Sayyid Khalid was allowed to return to Africa, where he lived quietly until his death in Mombasa in 1927.

An unfulfilled Irish interlude

During the year when the detainees from Zanzibar were attempting to extricate themselves from the island, a curious proposal emerged that, had it been enacted, would have seen them joined there by up to 4,000 dissident Irish Republican prisoners. These were men who had fallen into the custody of the Free State administration following the collapse of armed resistance to the treaty under which the Irish Free State had been constituted. The difficulties of coping with such a large and recalcitrant body of men at home formed the subject of a debate in the executive council of the Provisional Government in September 1922, which produced a resolution that the British government should be asked to house them in the remote fastness of its now famous (or infamous) prison-island of St Helena.[40] The implications were explored with some energy, leading to the drawing up of practical proposals and estimates of the likely costs. One such involved the uprooting of part of a First World War hutted camp in Derbyshire and its re-erection on the island; an alternative involved the construction of a purpose-built camp by the contracting firm run by Lieutenant Colonel P. N. Nissen, with accommodation in huts of the pattern that bore his name. Initial suggestions envisioning the establishment of the camp in some (unspecified) part of the island so remote as to require minimal supervision seem to imply that no one with firsthand knowledge of the topography was at hand to advise: rather than a circuit of barbed wire entanglements and a token guard, the British estimated a garrison cost of £200,000 per annum. The prohibitive expense of the proposed venture led to its early abandonment – perhaps just as well for the fledgeling Free State administration: as Paul Murray observes,

---

[39] Ibid., 167.
[40] This summary of the interlude is indebted to the account provided by Paul Murray (2003).

it is easy to imagine what its political opponents would have made of the imprisonment by an Irish government of thousands of Irish Republican prisoners, guarded by British troops and housed by a British contractor, on a remote British island to which they had been conveyed by British ships.[41]

Reactions on St Helena are unrecorded: no doubt nervousness at the impact that such a large number of recalcitrants might make on island society would have been mitigated by the prospect of a return to a buoyant prison economy, but quite possibly the entire chapter was opened and closed before word of it ever reached the islanders.

The Bahraini Three

On 22 December 1956 the *Government Gazette* of St Helena carried a most curious announcement:

> An urgent request made on behalf of Her Majesty's Government was recently received by His Excellency the Governor, as to the possibility of arranging for the detention in St Helena of five subjects of the Ruler of Bahrain in the Persian Gulf convicted of political offences. After discussing all aspects of this offence with the Executive Council, the Governor informed the Secretary of State for the Colonies of his concurrence in the proposed arrangement. It is expected that these persons will be brought to St Helena in one of Her Majesty's ships in the latter part of January, and that they will be detained at Munden's.

For the previous century Bahrain had been one of a number of British protectorates on the Persian Gulf, established by treaty but contested by increasing numbers of the population. It was in this context of political dissent, marked also by a dense web of dynastic disputes, that the ruler of Bahrain had arrested five men, all prominent members of the dissenting Committee of National Union which for the previous two years had promoted strikes, unrest and disturbances (not to mention demonstrations against the British action in Suez, which had finally precipitated the arrests).[42] They were now also accused of plotting to assassinate the ruler and his British hardline adviser, Sir Charles Belgrave; attempting to overthrow the government; attempting to depose the ruler; and organizing strikes and demonstrations. The Shaikh was anxious that three of the men – Abd 'Ali 'Aliwat, Abd al-Rahman al-Bakir and Abd al-Aziz al Shamlan – should be imprisoned by the British outside his territory, in order to neutralize their continuing influence. When the Political Resident pointed out the difficulty of the British government in concurring since the men had not actually been convicted of any crime, the Shaikh promptly convened a court (in which the three judges were all members of his own family) that quickly sentenced

[41] Ibid.
[42] The Bahraini National Union included many sympathizers with the Pan-Arab movement of Colonel Nasir in Egypt (not least Al-Bakir): for the background and much detail see Joyce 2000.

them to fourteen years' imprisonment, to be served in exile. It was at this point that St Helena had once again been seen as an ideal destination for them. Applying the conditions of the Colonial Prisoners Removal Act (1869), the British government made contact with the governor and on 27 January 1957 the prisoners arrived on the island on board the frigate HMS *Loch Insh*.

The men were duly installed in the former battery at Munden's Point, now encircled with barbed wire, and were placed under guard. Living quarters for them were found in a two-storey building known as Munden's House – thought to have been built as a guardhouse in the nineteenth century; the structure was now refurbished to form more comfortable living quarters; a few servants were provided to take care of their needs, but otherwise there was no intermingling with the island population. On a visit to Whitehall, Governor Alford reported that:

> prison conditions were good; the prisoners had access to all of the books and periodicals in the local library [and] a powerful short-wave radio, which received Middle Eastern stations. In addition, accompanied by a guard, once a week they were permitted jeep rides around the Island.[43]

Although the prisoners' trial and subsequent exile had been achieved in short order, things began to unravel in 1959 when one of them applied to the St Helena Supreme Court for a writ of *habeas corpus*, in which he challenged the legality of his imprisonment. A judge was brought from Nigeria to hear the case, attended by counsel from London and a Foreign Office adviser: ultimately the case was dismissed and a subsequent appeal to the Privy Council similarly failed, but having been aired so publicly the legitimacy of their detention had by now become a cause for considerable concern in Britain, culminating in a lengthy debate in parliament on 20 December 1960.

Every aspect of the case was closely questioned, with the government's position being repeatedly challenged: not the least crucial aspect was that the ruler of Bahrain seemed to have asked for an assurance that the prisoners 'will be sent to the island in accordance with the sentence decided' in a message dated 18 December 1956 – four days before the court had even been convened to lay charges against them and before the governor's announcement in St Helena. There were numerous other points to be addressed: no evidence for their defence was known to have been presented to the court – which in any case was hardly impartial – and there was indeed no formal criminal code operating in Bahrain at the time; the Colonial Prisoners Removal Act should never have been invoked since Bahrain was not a British colony and Britain had no obligation to become involved in the execution of Bahraini justice; the men had been removed from Bahrain to British jurisdiction (namely that of the captain of HMS *Loch Insh*) – without the necessary process of law being followed in England; had the facts come before parliament at the time, such involvement would never have been sanctioned. These matters were now of immediate interest since the agreement with Bahrain was said to run until the Shaikh should agree to the men's return to Bahrain;

---

[43] Quoted in Joyce 2000, 620.

he had now indicated his wish that they should do so and the government's position was that they were bound to comply; the opposition denied such an obligation existed and expressed the gravest fears for the men's fate if that were to happen.

Although the debate ended without a division, the climate clearly had changed very much in the men's favour, and when another of them similarly challenged the validity of their detention in June 1961 a court was again convened on St Helena, at which the appeal succeeded. In celebration, it is said that the men 'drank champagne, praised British justice and (still inside the prison) gave a luncheon attended by the Superintendent of Prisons, where they served Bahraini melons, grown in their tiny prison garden'.[44] In fact, Britain's role in the matter had been shown in no better light than that of the Shaikh of Bahrain and in time the men were compensated for the very dubious legality of their sentences. By that time, however, they had all been issued with St Helena passports and had left the island by the first available ship.

Asylum revisited

Napoleon's supplication to the Prince Regent in 1814 in which he presented himself as a worthy asylum seeker famously (or infamously) resulted in his being dispatched instead to imprisonment on St Helena. Reverberations from this two-centuries-old incident echoed through the press in 2020 when news leaked out that the Home Secretary of the day had instructed her civil servants to review the feasibility of using the island for the custody and processing of latter-day asylum seekers then illegally crossing the Channel in increasing numbers. Having earlier described the UK's asylum system as 'fundamentally broken' and faced with growing numbers of would-be immigrants detained while stowing away on lorries or plucked from the sea as they committed themselves to increasingly desperate Channel crossings in unseaworthy dinghies, the Home Secretary was looking for radical ways to fix things.

When the story broke in the *Financial Times* on 29 September 2020 it caused something of a political furore. It transpired that a Whitehall brainstorming session had been convened to consider any and all options on how to tackle the problem. To prevent the migrants from simply disappearing into society, the meeting considered (among other possible solutions) the feasibility of 'offshoring' them, either in decommissioned ships, on one or other of the islands around the British coast – or even as far away as the mid-Atlantic. A precedent for such a solution was held out in the practice adopted by the Australian government of housing migrants trying to reach their shores from South-East Asia on the distant island of Nauru and offering very minimal chances that they would ever be allowed to reach their intended destination. When confronted with the leaked story that St Helena and Ascension Island were being considered as holding areas while the individual migrants' claims for asylum were processed, the British government maintained that it was simply exploring all options, even 'implausible ones', but such a bland

---

[44] Joyce 2000, 622.

Later detainees, 1800s and 1900s

response was never going to avert the brickbats from political opponents inside parliament and beyond: the Shadow Home Secretary characterized 'this ludicrous idea' as 'inhumane, completely impractical and wildly expensive'; a spokesman for the Scottish National Party concurred that the government's treatment of asylum seekers was 'utterly toxic' and the fact that it had even considered shipping refugees thousands of miles to remote islands in the South Atlantic, 'like some sort of modern day penal colony', brought shame on the UK and was 'out of step with the values of equality, community, justice and human rights'. The Chatham House think-tank characterized the whole concept of 'externalization', designed to exclude asylum seekers, as undermining international protocols for the handling of refugees. Official claims that the suggestion had emerged merely from blue-skies thinking were countered by press reports that the government had been working for weeks on the scheme and 'scoping everything', including costed estimates for building detention camps on Ascension and St Helena.[45] There was much embarrassment all round.

However long its gestation, the idea could scarcely survive the derision poured on the heads of ministers, leading the government to quietly shelve the idea. It remains fascinating that St Helena, having once held Europe's most famous exile and periodically afterwards some dozens or hundreds of others, should continue to resonate so powerfully in the corridors of Whitehall as a byword for remoteness and impenetrable security whenever the political need arises for moving a sensitive problem as far away as possible.

---

[45] See, for example, *Financial Times*, 29 September 2020. Meanwhile, 4,000 miles away in the mid-Atlantic, Alan Nicholls, a member of the Ascension Island Council, dismissed the possibility as a 'logistical nightmare', adding that on Ascension the security problems raised by the presence of two military bases on the island could render it prohibitively expensive.

9

A place in the modern world

It is a commonplace that the world became much smaller for everyone in the course of the twentieth century, but for the inhabitants of remote St Helena the changes took longer to materialize than for most. It might be mentioned that while the logistical and technological changes that form the basis of this final chapter were being acted out, the national status of the islanders themselves within the world underwent a period of unaccustomed turmoil.

Having formerly enjoyed undifferentiated full British citizenship since the granting of the East India Company's royal charter of 1673, St Helena was reclassified as a British Dependency in 1834.[1] For a century and a half the change of status had little discernible impact, but with the introduction of the British Nationality Act 1981 the islanders found themselves suddenly placed at a major disadvantage when (along with thirteen other such territories) they were deprived of their rights of residence in the UK – collateral casualties of government moves to block potential large-scale immigration from Hong Kong to the UK as Britain's ninety-nine-year lease on the New Territories approached its end. Although the numbers directly affected on St Helena were comparatively small, the slight was deeply felt by the entire population, which had come to regard itself as occupying 'the lost county of England'. Two decades of intensive lobbying followed, with the islanders taking their complaint of injustice as far as the United Nations. Eventually, with the passing of the British Overseas Territories Act on 21 May 2002 (the putative 500th anniversary of the island's discovery), full rights were restored. Even with its national status cemented, far-reaching changes in St Helena's international relations continued to unfold, as they had done throughout the island's history.

The coming of the submarine telegraph

The Boer War had brought to the island a more oblique benefit than that represented by the prisoner-of-war camp (Chapter 8): this was the arrival on 26 November 1899 of a submarine telegraph cable, laid with the primary aim of

---

[1] St Helena itself acquired Ascension Island as an administrative dependency in 1922 and Tristan da Cunha likewise in 1938.

A place in the modern world

Fig. 33   Cable-laying Ship *Anglia*. Postage stamp issued to mark the centenary of Cable & Wireless, 1999.

enhancing communication between South Africa and London. The island was to form the first station on a route that would take the Eastern Telegraph Company's 1½-inch copper cable from Cape Town to St Helena, onwards to Ascension Island and thence to St Vincent in the Cape Verde Islands; from there the signal was to be carried to Madeira and mainland Portugal before finally coming ashore at the Porthcurno telegraph station in Cornwall.[2] The stretch from Cape Town to St Helena was laid by the Telegraph Construction and Maintenance Company's cable ship *Anglia*, an achievement whose centenary was celebrated in 1999 with a commemorative postage stamp (Figure 33). The *Anglia* was a modern steamship, but a report from the time is illuminating in showing the continuing importance of sail at the end of the nineteenth century in St Helena's daily life and the exceptional significance of the telegraph to the merchant mariners:

> The connection by cable with the 'mother country' will be the greatest boon ever conferred upon St Helena, for the reason that hundreds of merchant sailing ships from the East, which under existing arrangements go out of their way to call at the Brazils, Cape Verde, and the West Indies for orders, will avail themselves of this station, St. Helena being situated in the heart of the southeast trade winds, and consequently in the direct track of ships from India and China bound to the continents of Europe and America.[3]

The oceanographic advantage that had proved critical throughout the island's development continued to play a crucial role in its economy. The new cable came ashore at Rupert's Bay, where Eastern Telegraph established a cable station; administrative offices were also created at The Briars – earlier the scene of Napoleon's introduction to incarceration on the island.

Just ten years short of the centenary, another milestone in the island's communication history was marked by the construction of a satellite ground

---

[2]   See 'History of the Atlantic Cable & Undersea Communications' at https://atlantic-cable.com/Cables/1899StHelena/index.htm.

[3]   Report from the American consul on the island, R. P. Pooley, in *Commercial Relations of the United States with Foreign States during the year 1898* (Washington, DC, 1899), 250.

station at The Briars (by now owned by Cable & Wireless), its 25-foot satellite dish providing an international link through Intelsat 707 to Ascension Island and the UK. Since 2015 a cellular telephone network has linked almost the whole population and has provided a full range of voice calls, text messaging and mobile data.[4]

With the new millennium the limitations of the 100-year-old submarine cable had begun to become increasingly apparent. A campaign was mounted to connect the island to the planned South Atlantic Express cable, linking South Africa and the Americas, with a view to providing high-speed internet access to the outside world: agreements were drawn up but the project failed to materialize. On 29 August 2021 Google's Equiano submarine cable – like its ageing predecessor connecting Portugal and South Africa via the Canary Islands but with twenty times the capacity – was landed at Rupert's Bay. The 700-mile-long spur from the principal cable was commissioned in 2022,[5] providing a state-of-the-art fibre-optic connection that will enable everyone on the island to access on an equal footing the global networks on which the world has come to depend for so many aspects of everyday life, with education, health and business highlighted as of primary importance.

Radio and television

Public radio came to St Helena only during the course of the 1960s, when plans hatched by one of the islanders in collaboration with two engineers from the Diplomatic Wireless Service resulted in the establishment in 1967 of a local station, Radio St Helena, which provided a service to the islanders for the following forty-five years. The station (officially the St Helena Government Broadcasting Station) broadcast a medium-wave AM service from transmitters initially sited at Longwood farmhouse and later in a higher location in St Paul's, with a range of some sixty miles. Initially the 500 watt transmitter drained the locally available electricity supply to such a degree that houses in the area would periodically lose power, but in time the supply was increased while the transmitters were doubled in size, offering a diet of news, music and educational programmes. Programming was initially limited to a few hours each week but was gradually extended; by 2000 a full day's schedule was achieved, seven days a week. From 1990 a modest outside broadcast unit was set up, and once a year the station extended to international coverage on short-wave, marking 'Radio St Helena Day' by broadcasting to audiences around the world, until the short-wave antennae were destroyed in a storm.

---

4   A land-based telephone system for internal communication on the island had been introduced in 1886 (Denholm n.d., 36–40).
5   https://www.datacenterdynamics.com/en/news/google-officially-launches-equiano-subsea-cable/

A place in the modern world

A new and privately owned station, Saint FM, opened in 2005, its signal reaching parts of the island inaccessible to Radio St Helena. Both stations closed in 2012; the following year Saint FM Community Radio – corporately owned by members drawn from the island's population – took over the channels vacated by its predecessors. Television services are provided to the island by Sure South Atlantic Ltd, with fifteen channels available by subscription. Most of the content is of British origin, made available through South African MultiChoice by means of a satellite dish at Bryant's Beacon from Intelsat 7.

Maritime oblivion again

Within five centuries of its discovery in the remote mid-Atlantic, St Helena's continuing prominence on the nautical route map entered a more precarious phase following World War II. For a time it continued to enjoy the passing glamour of stopovers from ships of the Union-Castle Line, carrying holidaymakers, emigrants and businessmen in some style on the long voyage to Australia, but even that privilege was extinguished when Union-Castle withdrew from the route in 1977, in the face of dwindling profitability.

The origins of the island's struggle to maintain its place on the international navigation charts can be traced back to the early years of the previous century, when the EIC had sent a 136-ton armed schooner – the *St Helena* – specially commissioned at Blackwall, to carry stores and livestock from the Cape to the island. From 1814, under the control of the government of St Helena, she attended visiting East Indiamen and made occasional voyages as far as Mauritius and Rio de Janeiro, but her regular routine had her shuttling to the Cape and back until 1830. In that year she was returning to England for repairs when she was captured in the Gulf of Guinea by pirates who killed thirteen of the ship's complement, cut away the masts and tried unsuccessfully to sink her; miraculously, the survivors among the crew managed to sail the *St Helena* to Sierra Leone, whence she was returned to England – surviving capture by the Portuguese on the voyage – only to be sold. Thereafter the EIC supplied the island by means of contracts let to shipping companies at the Cape.

Although not the first time that steam had appeared on the horizon, the unexpected arrival at St Helena in September 1852 of the first iron steamship to cross the Atlantic – Brunel's massive SS *Great Britain* – must have brought home to those who witnessed the event the precarious future of this community, forged in the days of sail.[6] The ship had been on her first direct voyage to Australia via the

---

[6] *Great Britain* did indeed have secondary masts to supplement the steam power, and after thirty years on the Australia run was converted to all-sail in 1881. After carrying coal and wheat between the UK and the USA she was damaged rounding Cape Horn in 1886 and sought shelter in the Falkland Islands; the cost of repair proving prohibitive, she was sold to the Falkland Islands Company and served as a coal hulk. A

Cape when she began to run short of fuel after battling headwinds for much of the voyage along the west coast of Africa; the captain decided he had no option but to divert to St Helena, then 1,000 miles north-west of his current position. The island had had a coal depot since the 1830s, established under Governor Dallas (though extensively damaged by rollers in 1846), but had quickly gained a reputation for high charges that led many masters to steer clear whenever possible.

Although by 1852 St Helena had become a regular port of call for vessels of the General Screw Steam Shipping Company, the fact that *Great Britain* arrived in need of 500 tons of coal and left a week later with only 100 tons from shore and 100 tons procured from a Royal Navy ship, supplemented by several hundred tons of firewood, suggests that all was not yet fully prepared for the new era.[7] A further significant blow to the island's maritime economy took place with the opening of the Suez Canal in 1869, which greatly diminished the amount of seaborne traffic on the Atlantic route to the east and contributed further to St Helena's isolation.[8]

With the transfer of the island to the Crown in 1834 and with the rise of steam-driven transport, St Helena came to rely for communication with the outside world on the services of the Union Steamship Company, founded in 1853 and awarded a mail contract to South Africa four years later. In 1876 a second carrier, the Castle Mail Packets Company, was awarded a second contract; sailing on alternate weeks and calling at St Helena on both the outward and return voyages, the two companies provided a remarkably regular service until they merged to form the Union-Castle Mail Steamship Company in 1900.[9] Thereafter a more intermittent passenger service was provided, latterly by cargo-liners known locally as 'two-hour ships' on account of the tight turn-round time allowed at the island. Ultimately competition from direct air travel from Britain to South Africa and Australia led to the company's demise in 1977.

The loss of visits from the Union-Castle Line brought about more than increased cultural isolation for the island, for as well as postal communications the liners

---

rescue mission in 1973 managed to refloat her, and she was towed back to Bristol to be conserved and to serve as a major tourist attraction.

[7] See the chapter 'St Helena as a coaling station: a summary of evidence', in Hearl 2013, 293–301. The necessity of loading and unloading coal by lighters at Jamestown inevitably added to its expense there. Steven Gray (2018, 26–29) suggests that it was only in the late 1870s that the strategic significance of steam power and the necessity for a supporting network of coaling stations became fully recognized, but St Helena never became a primary station.

[8] G. C. Kitching, in a note dated 1947 and added to his *Handbook and Gazetteer* of ten years earlier, asserts that 'it was not the Suez Canal that killed St Helena ... Towards the end of 1860, the Crown suddenly removed all its establishments, and then came the opening of the Suez Canal. From 1873 onwards Governors were powerless. They had no money and no staff.'

[9] McCracken (2022, 185) notes that in 1880 mail ships called at the island thirteen times a year southbound and twenty-six times northbound, the former being subsidized by the British government and the latter by the government of St Helena.

had carried important (if small-scale) consignments of freight. Recognizing the serious implications for the island's economy, the British government responded by purchasing a vessel that would carry passengers and essential goods, as well as the mail, on a regular, predictable and dedicated service: the new Royal Mail Ship would again be named *St Helena*. The ultimate arrival of 'The RMS', as she was locally known, returned the island's communications to a regular and locally controlled footing for the first time in several generations.

Built in 1963 and originally sailing as the *Northland Prince* on a service between Vancouver and Alaska, the vessel was recommissioned in 1978 by the newly formed St Helena Shipping Company (Plate 27). At 3,150 tons, the *St Helena* successfully served the needs of the islanders for more than a decade, offering an appropriate mix of berths for seventy-six cabin passengers (who continued to be hosted in the style of the now-defunct liners) and camp beds on deck under an awning for the more indigent, together with roomy cargo space. Her normal routine took her six times a year from Avonmouth to Cape Town, calling initially at Las Palmas, Ascension Island and St Helena on the way, with Santa Cruz de Tenerife later replacing Las Palmas. For islanders the Canary Islands stop was of particular importance, allowing the ship to fill up with fresh fruit and vegetables to supplement the local produce. The northbound voyage followed the same route, with occasional stops at Vigo.

A wholly unexpected interruption to this established pattern was brought about by events on the other side of the Atlantic, when the *St Helena* was requisitioned in 1983 as one of the 'ships taken up from trade' to serve in the aftermath of the Falklands War. Fitted with pipelines for refuelling at sea, a helicopter deck and four Oerlikon 20 mm cannon for defence, the *St Helena* served for some months as a support ship for a small minesweeping squadron comprising HMS *Ledbury*, HMS *Brecon* and five converted trawlers. Nineteen of her St Helenian crew remained on board as volunteers and all returned safely in the company of the two minesweepers in late August (Figure 34); the *St Helena* would remain under charter to the Ministry of Defence for a further year.

The island meanwhile was served by chartered ships, and by the time the *St Helena* returned to her more usual routine it had become quite evident that the volume of traffic to and from the island had increased to the point where a larger vessel was required. A new RMS was commissioned from the Aberdeen shipbuilders Hall Russell & Co. but within months of her keel being laid the company had become insolvent. After facing the prospect of being broken up before she had been launched, construction of the *St Helena* was completed by A. & P. Appledore; despite spiralling costs she at last made her maiden voyage in 1990. At 6,767 tons and operating with a crew of fifty-seven, the new vessel became a popular icon and indeed functioned as a seagoing ambassador for the island itself (Plate 28): the complement of passengers was doubled at 128 berths (increased in 2012 by a further twenty-four), catered for with a swimming pool, gym, medical facilities and several lounges; deck passengers were no longer

Fig. 34 RMS *St Helena* executing replenishment at sea with Hunt Class minesweepers HMS *Ledbury* and HMS *Brecon*, 1983–4.

carried. Cargo capacity was one-third greater than that of her predecessor; heavy-lifting cranes enabled easy loading of up to ninety-two 20-foot containers, while space was provided for motor vehicles and livestock, and for oil products. Her routine itinerary to South Africa and Ascension Island was broken from time to time by voyages to England. On the return leg from one of these, in November 1999, the vessel broke down and had to be taken into harbour in Brest for repairs; the following year she suffered an engine-room fire at sea between Cardiff and Tenerife on her way back to the island, while later propeller problems led to the cancellation of several voyages – all leading to the stranding of passengers and a certain amount of panic-buying among islanders faced with delays of supplies of uncertain duration. In June 2001 the *St Helena* made a round trip to Tristan da Cunha, some 1,500 miles to the south and normally visited only once a year by the RMS, to deliver essential building materials for making good the devastating hurricane damage inflicted on the island the previous month.

From 2011 regular services to England were discontinued, with Cape Town, St Helena and Ascension Island now forming the ship's normal itinerary. Minds were once again focused on the vulnerability of the island's supply line, dependent on a single ship; calls intensified for the establishment of an airport that would provide the means for more frequent and dependable communication. While the airport itself suffered serial problems (see below) the *St Helena* was kept busy ferrying men and materials for the new project – not to mention larger-scale items such as the wind turbines that now provide the island with much of its electricity – and catering for a modest flow of tourists on the five-day voyage from Cape Town.

After more than one reprieve in order to substitute for the overrunning airport project, *St Helena* eventually retired from service with her last voyage from the island on 10 February 2018. Remarkably, her funnel still displayed the device of the East India Company – incorporated (with local acclaim) into the emblem of the St Helena Shipping Company – acknowledging the long-past era in which the island had found itself at a crucial nexus in the east-west trading network.

The end of one role for the ship was quickly followed by another, for in April of the same year she was recommissioned as an anti-piracy 'vessel-based armoury' in the Gulf of Oman, providing security equipment and personnel for vessels passing through the potentially dangerous waters of the Strait of Hormuz, and the following October was sold again. In 2020 she completed an extensive refit to become a mobile hub for the Extreme E electric car racing initiative: personnel and equipment are carried to the series of remote and environmentally challenging venues where races are scheduled to take place, with the vessel functioning as a floating garage and control centre while also providing living and recreational accommodation. With her engines and turbines extensively rebuilt and re-engineered for minimal environmental impact, the much-loved *St Helena* looks set for a glamorous new lease of life at the cutting edge of automotive technology.

Although the scheduled passenger service that the islanders considered their own has probably gone for good, freight continues to arrive by sea by means of container ships – notably the MV *Helena*, which provided a sole lifeline to the island when the air link with South Africa was cut during the Covid crisis. The rising popularity of cruising holidays promises to bring short-term visits to the island from potentially large numbers of holidaymakers. With capacity for many hundreds of guests, these large vessels rarely anchor at St Helena for more than a night or two at a time, each disgorging their passengers with a minimal amount of shore-time to explore the island and inevitably falling into well-established itineraries that allow for visits to a handful of high-profile sites but little more. Their impact on the economy, while welcome, is presently quite narrowly focused; shops may extend their opening hours (though not all benefit noticeably) and no doubt much ingenuity will be brought to bear on spreading the benefits more widely. Already the advantages of a newly constructed dock at Rupert's Bay – the first in the island's history – are being felt in improved accessibility for tourists: rough seas during the summer season (when the majority of cruises arrive) had all too often prevented passengers from disembarking, causing loss of revenue to the islanders and leading some shipping companies to omit the island altogether from their itineraries. Now, while the ships still anchor in the roads of James Bay as they have done since the sixteenth century, their tenders can come and go with an ease that would have been the envy of early mariners.

However these visitors arrive, it seems certain that the future viability of the island's economy will be bound up with the tourist industry. A Land Development Control Plan drawn up in 2007 states bluntly that 'the development of tourism has a major part to play in securing a prosperous future for St Helena' – indeed it finds that 'the growth of tourism is crucial to the island's economic recovery'.[10] After a period when ever-larger cruise ships were launched on an annual basis, the experience of the past few years has shown how vulnerable to disruption the entire sector has been, millions of tons of shipping having been mothballed around the world in response to Covid-19. St Helena felt the full impact of that disaster (even while ensuring that not a single case ever reached the island), but the last decade has also seen the realization of an alternative form of access – one that has been keenly anticipated for the past hundred years.

Birth of an airport

From the time the Wright brothers took to the air at the opening of the twentieth century, there must have been many who speculated about the possibility that St Helena might one day be opened up to air travel. Even allowing for the remoteness of the island, few would have imagined it would take over a century for that dream to become a reality.

As so often happens, the exigencies of warfare provided a powerful catalyst. In 1943 a survey party from the South African Air Force spent a few weeks on the

---

[10] St Helena Government n.d., 7-4.

island, assessing its suitability as a base for South Atlantic patrols, to extend the range of those flown from the South African mainland. They quickly identified the plain at Prosperous Bay as a possible site for a runway, but concluded that, while feasible, it did not then represent a practical proposition. The idea was revisited in the 1960s, and eventually in the 1980s a company named St Helena Airways was formed with a view to putting the plans into action: a new survey was carried out and in 1985 a low-level fly-by by a Hercules aircraft of the RAF tested the aerial conditions, but another decade would pass before the proposals would gain traction with the St Helena government. In Whitehall too, interest in the southern waters had been piqued by the experience of the Falklands dispute, which had unfolded some 3,800 miles to the west, and the idea of an air link at last began to gain official backing. In 2002 a referendum held on the island resulted in three-quarters of the population voting in favour of developing an airport rather than updating the RMS service, and three years later the British government declared its support for the scheme: an airport would indeed be built, scheduled for completion in 2010 and funded by the Department for International Development.[11]

There were difficulties from the beginning. The Prosperous Bay plain again emerged as the favoured location, but a review produced in 2004 had shown that a number of endemic plant species, lichens and invertebrates depended critically on the preservation of the plain's ecosystem;[12] it is also an important nesting area for the island's iconic (though critically endangered) Wirebird.[13] In response, strategies to mitigate the impact of the airport were developed, designed to enhance the habitat available to native species while removing those classed as intruders. The ironically imperialist echoes of the situation, in which one of the major promoters of the scheme, the St Helena Leisure Corporation (Shelco), had as its co-founder a former chairman of the Campaign to Protect Rural England, were not lost on the population.

Then all the companies initially tendering for contracts to build and operate the airport withdrew, citing as their reason the non-cooperation of the British government in allowing for preliminary on-site investigations before the submission of costed estimates. Plans were in any case placed in suspension as a

[11] United Press International quoted analysts as stressing the continuing stand-off with Argentina as a significant factor in the decision, whereas the St Helena authorities maintained the airport's potential importance for the development of tourism and providing an easier transport link for islanders who had found work elsewhere. Imaginative alternatives considered included a smaller airstrip serving a hub to be developed on Ascension Island, water aerodromes serving amphibious aircraft operating on the same route, and long-range airships flying to Cape Town (St Helena Government n.d., 2-2).

[12] Myrtle Ashmole (2005, 27) reported that the plain contained 'an extraordinary concentration of endemics – over 50 have been found here and 35 of these occur in this limited area and nowhere else on St Helena, let alone elsewhere in the world'.

[13] See the revised Wirebird Species Action Plan (2011) at https://sainthelenaisland.info/wirebirdspeciesactionplanrev2011.pdf.

result of the severe economic downturn in 2008 and ultimately it was only in 2011 that contracts were finally agreed; preparatory work on site began the following year. Plans for the construction of a luxury hotel and resort which surfaced in the course of the planning process failed to materialize, but an incidental benefit from the construction process was the provision of a jetty, built in 2012 at Rupert's Bay. Mentioned above in relation to the tourist trade, this had a more immediate role in facilitating the unloading of materials and heavy machinery for use in the airport construction – the like (and scale) of which had never before been seen on St Helena. Enormous amounts of materials and machinery had to be transported to the island from Walvis Bay in Namibia; a specially chartered heavy-lifting vessel shuttled back and forth on a three-week cycle to supply everything from fuel to high explosives and massive trucks and excavators to undertake the civil engineering work – not the least of which involved transporting everything via the difficult nine-mile land route to Prosperous Bay.

The logistics involved in constructing the specified 6,400-foot runway were daunting by any standards. To begin with, a deep chasm, Dry Gut, lay in the path and needed to be filled-in to a depth of almost 330 feet: it would be the summer of 2015 before the last of the 280 million cubic feet of ballast required had been emptied into the gorge and construction could proceed. A wide embankment underpinned the runway and an 800-foot safety area extended the southern end to accommodate any unscheduled 'excursions'. From 2015 a number of test landings took place, in which a variety of aircraft probed the airport's operational suitability and helped calibrate its instrument landing systems.

At this point the scale of the difficulties imposed by its location began to emerge. The problem came from the airstream sweeping down from the moist upland regions in the form of so-called 'fall-winds'. Their moisture content causes them to sink close to ground level just as they encounter the runway, where they meet with the updraught from the steep coastline of Prosperous Bay: the result is a phenomenon known as wind shear – just about the most difficult form of turbulence a pilot can encounter in taking off and, more particularly, in landing. The problem should have surprised no one: in essence (though not in detail), the phenomenon had been remarked upon by Charles Darwin, while walking near the Asses Ears in the south of the island almost 200 years earlier:

> One day I noticed a curious circumstance: standing on the edge of a plain, terminated by a great cliff of about a thousand feet in depth, I saw at the distance of a few yards right to windward, some tern struggling against a very strong breeze, whilst, where I stood the air was quite calm. Approaching close to the brink, where the current seemed to be deflected upwards from the face of the cliff, I stretched out my arm, and immediately felt the full force of the wind; an invisible barrier, two yards in width, separated perfectly calm air from a strong blast.[14]

---

[14] Darwin 1860, 491.

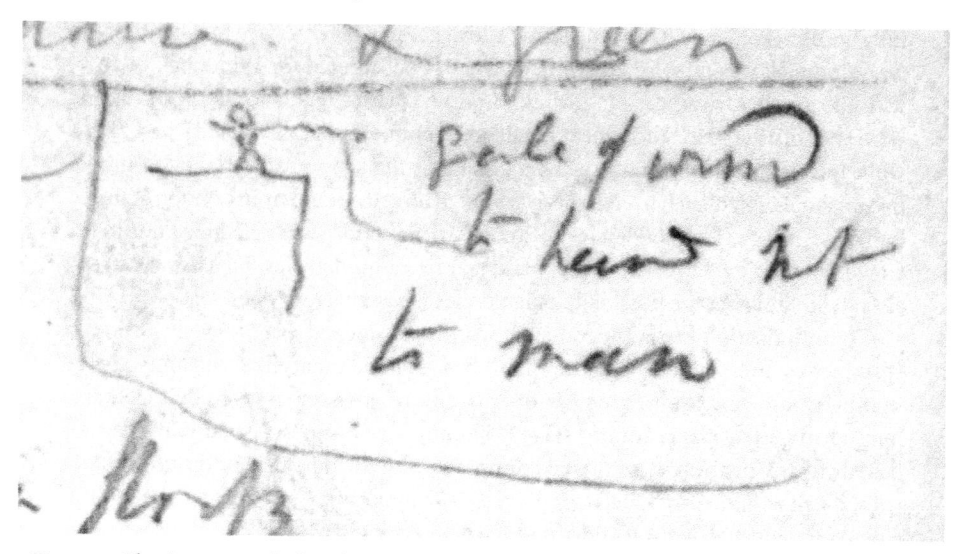

Fig. 35　Charles Darwin's sketch of the updraught felt at the cliff-edge, 1836. The annotation reads 'Gale of wind to hand not to man'.

A crude sketch in his notebook illustrated the point (Figure 35).

The updraught from cliff faces is routinely exploited by soaring birds and hang-gliders, but the consequences of its intersecting here at right angles with the offshore wind were potentially lethal for aircraft. Islanders had the mortification of seeing their brand new airport stand idle for a year and a half while the authorities came to terms with the implications. The need for additional windspeed indicators to provide real-time information on wind conditions was recognized, but although the airport received its official Aerodrome Certificate in May 2016, it failed to receive a Civil Aviation Authority Safety Certificate clearing it for commercial operation until 5 October 2018.[15] By this time the company that had won the contract to operate the airport had gone into liquidation; consequently, its administration was now taken over by the St Helena government.[16]

Although the runway had now been declared usable, traffic was limited to comparatively small (seventy-seater) aircraft flown exclusively by experienced pilots specially trained for operating from the island, where the need to land with a tailwind is common;[17] the airport nonetheless operates an open skies policy that permits any

---

[15] Issues over the monitoring of turbulence and wind shear were the principal factors in the delay.
[16] In addition to the runway there are all the usual facilities for air traffic control (including an instrument landing system), refuelling, and passenger-handling in an efficient terminal building.
[17] The frequent turbulence at St Helena prevents larger aircraft taking off on a northerly course, but the option of landing in either direction is necessary for safe operation at the airport.

fully licensed company to operate there. Ultimately a weekly service to Johannesburg via Windhoek was established (with Walvis Bay substituted for Windhoek in 2019) with the South African carrier Airlink; monthly charter flights to Ascension Island were later instituted for the benefit of islanders employed on the military base there, while interest in establishing longer-range charter flights was expressed by a number of companies. Regular charter flights were flown directly to and from London Stansted by Titan Airways (Plate 29) while the South African service was suspended due to the Covid pandemic in 2020–21 – supplemented by cargo flights in increasingly heavy aircraft, holding out the possibility of larger-scale passenger services.

Communication with the outside world has indeed been eased, although sustained economic benefit to the island has remained marginal; certainly visitor numbers have come nowhere near the maximum of 58,000 a year envisaged as a long-term goal at the planning stage.[18] Numbers of businesses that had invested heavily in anticipation of increased commerce with the outside world were already ruined when the airport failed to deliver significant traffic – and that was before the Covid epidemic wrought its devastating effect on international travel. Progress in the economic sector has still to be detected, but there were other ways in which the airport was expected to contribute to the island's continuing development, such as improved access to tertiary education for the young, more effective health care facilities and even the creation of an economic climate in which the population might contemplate the benefits of self-determination. In looking forward to the wide range of benefits that are foreseen, the airport's teething problems may represent only minor set-backs in a long but still potentially rewarding process.

By way of an interesting aside, and with an eye on the long history of scientific preoccupation with variations in the magnetic field on the island, an amendment had to be made to the designation of the airport's runways within four years of their opening to traffic. In February 2022 runways 20 and 02 were redesignated 19 and 01, registering a one-degree shift in magnetic north that had become apparent in a remarkably short space of time. Although the geophysical explanations have become immensely more sophisticated in the past 200 years, present-day navigators find themselves having to take account of phenomena that would have been very familiar to generations of their forerunners.

Biosecurity and regeneration in the new international age

Particular attention has been paid in earlier chapters to the extensive impact made on the face of St Helena by human interventions, both deliberate and accidental. The enhanced international accessibility promised by the airport, as also by visiting cruise ships, carries with it the unwelcome possibility of opening up a new frontier for the introduction of pests, diseases and alien species, against which a new agency has been established – Biosecurity St Helena. A number of specific threats have been

[18] St Helena Government n.d., 5-8.

identified and controls put in place. Plants, fruit and flowers may be imported only under licence and seeds brought into the island must be commercially produced and packaged; the importation of honey has been completely prohibited, in an attempt to avoid the introduction of diseases that might threaten the island's bee population and its own expanding honey industry; some meat and fish products are also controlled.

Other agencies are hard at work aiming not only to control and stabilize the island's ecology but, wherever possible, to reverse losses suffered in the past in order to reclaim something of the diversity of habitat that characterized the island at the time of its discovery. Data from the Sustainable Cloud Forest Restoration Project feeds into initiatives for habitat management, disaster emergency planning, sustainable public water supply and climate change adaptation.[19] Work on controlling invasive mammals such as rats and feral cats – the formerly ubiquitous goats have been eliminated from the wild within the past few decades – goes hand-in-hand with campaigns to stem the proliferation of New Zealand flax (see Plate 11) and blackberries while re-establishing viable populations of native plants. A special unit within the Agriculture and Forestry Department of the St Helena government (established in 1934) is tasked with rehabilitation of the impoverished forest landscape, promoting the recovery of endemic populations wherever possible while also developing economically viable and ecologically sustainable forest resources. Natural stands of gumwoods and other native species have been multiplied by means of cuttings and seeds nurtured in plant nurseries, with care being taken to maximize genetic diversity within the much-reduced gene pool. By 1988 plans were in hand to double the extent of the designated national forests to some 9,000 acres, representing about 30 per cent of the island's surface.[20] The initiative has already achieved some success with the establishment of a Millennium Gumwood Forest Project in the north-eastern corner of the island (Plate 30), where some 250 hectares of woodland have already been established,[21] forming an inspiring example of what may be achieved by similar initiatives. In so many ways, St Helena continues as a microcosm of the wider world.

Today initiatives of this kind are coordinated and monitored under the aegis of the St Helena Research Institute, which has a mission to advance learning related to the island and to ensure that the results of new research are made accessible and beneficial to the islanders themselves. It also operates within the network of similar organizations that make up the South Atlantic Research Institute, ensuring that its contributions are registered at a wider regional level.

---

[19] At a fundamental level, the St Helena Climate Change and Drought Warning Network is focused on understanding and ameliorating the water-shortages – often amounting to severe drought – experienced over the centuries and occurring with increasing frequency in recent years.

[20] Barlow 1989, 62.

[21] An estimated 55,000 trees will be required to fulfil the planting scheduled for the project area: https://sthelenaisland.info/the-millennium-forest/

St Helena: An Island Biography

And next: Utopia in the twenty-first century

The potential for further impact beyond the island's shores has been highlighted in recent years by international interest in exploration of the ocean floors as possible sources of mineral wealth. Rising oil prices at the turn of the present century led to a scramble among the former colonial powers to establish their rights over the territorial waters surrounding their surviving overseas possessions and to register them with the United Nations by 2009, after which point no further applications would be deemed valid. As a result, tiny St Helena found itself at the centre of a mid-Atlantic zone, potentially of a 200 to 350-mile radius, within which the British government claimed exclusive rights to explore the seabed with a view to extracting such resources as it might contain. (Ascension Island, Tristan da Cunha and the Falkland Islands were similarly zoned.)[22] Mineral nodules lying on the deep ocean floor provided one area of interest, but oil was uppermost in the minds of the contenders. Formerly shielded by their extreme depth, advances in drilling technology were beginning to bring any such resources at least potentially within the reach of the engineers. Soon these possibilities – largely hypothetical at present – will become realities: it would take a bold person to foresee where they might take St Helena in the coming decades.

The progressive loss of St Helena's virgin innocence over five centuries of destructive encounter with humanity has been outlined here. It would make depressing reading – particularly given the island's characterization as an analogue for much of the damage wrought throughout the natural world by human impact during the colonial and post-colonial era – were it not for the redemptive efforts of those who continue to strive to mitigate and to reverse these negative trends. The historical record highlights several occasions at which perceptive observers grasped the significance of the environmental degradation all too obvious before their eyes and the bleak consequences of continuing failure to address the root causes. Well-meaning initiatives aimed at stemming one aspect or another of the island's environmental decline emerged from time to time, but all seem to have evaporated within the space of a few years, victims of frequent changes of administration or indeed conflicting agendas followed by the respective administrations on St Helena and in London.

Once it had become enmeshed in the web of mercantile and geopolitical interest that spanned the world with increasing complexity from the decades around 1500 – and it may be remembered that St Helena was discovered within ten years of Columbus's symbolic opening of European contact with the New World – the island's fate became inextricably linked with developments taking place far from its shores. Although never able to exert influence on these processes, the island's strategic importance remained considerable, whether in its crucial role in

---

[22] See Ambrose Evans-Pritchard at http://www.telegraph.co.uk/finance/comment/ambroseevans_pritchard/2790589/Oil-crisis-triggers-fevered-scramble-for-the-worlds-seabed.html.

facilitating trade with India and the Far East, as a way-station in the trans-Atlantic slave trade and later in the suppression of the same diabolical business, or as an acclimatization station for Oriental plants destined for Britain or – in the case of those implicated in Sir Joseph Banks's ambitions for a global exchange network – for the West Indies. Without St Helena, Britain's empire in the East could scarcely have taken shape: Jacob Bosanquet, an East India Company director, aptly characterized the island as 'the principal link of that chain which connects this country with her Indian possessions'.

This perceptive observation might stand to presage a whole emergent field in present-day scholarship within which 'island studies' have come to form a subject area in their own right, focusing on the role of islands as 'vibrant entrepôts sitting at the heart of complex oceanic networks that linked peoples, polities and cultures'; rather than forming subjects for individual study, islands offer here 'methodological spaces for interrogating the impacts and legacies of globalizing forces, such as imperialism, migration, and long-distance oceanic trade'.[23] The importance of St Helena as a key node in these networks will ensure its continuing interest in this field, even if the island ever struggled to become economically 'vibrant' in its own right. Today its strategic importance is negligible, yet there is little inclination on the part of the British government – or of the island's population – to contemplate the possibility of mutual disengagement. All the signs are, indeed, that any such distancing would only exacerbate the prospects of a population already dwindling in numbers, increasing in average age and still lacking the long-sought-after magic formula that might lead to self-sustainability.

The passage of time gives us the advantage of seeing from a detached viewpoint not only the island's dystopian past but also the potentials and difficulties that lie in its future. The challenges facing this 'speck in the ocean' now, no less than those it has contended with historically, are indeed best understood against a global perspective, even if the capacities of the islanders to exert significant influence on their fate are no stronger now than they ever have been. The local initiatives now under way (themselves conceived on an international scale), however, have already produced positive results: like William Blake's Jerusalem built in England's green and pleasant land, who is to say that Utopia may not yet find itself re-established on St Helena?

---

[23] Hamilton and McAleer 2021, 3.

Appendix: Governors of St Helena

Since historical eras on St Helena are commonly referred to in the text (as elsewhere) as 'during the governorship of ...', it may be useful to summarize here the dates of their respective periods of office. For these purposes, no account is taken of the periods of office held by deputy or acting governors.

East India Company

Captain John Dutton 1659–61
Captain Robert Stringer 1661–71
Captain Richard Coney 1671–72
Captain Anthony Beale 1672–73
Captain Richard Munden 1673
Captain Richard Keigwin 1673–74
Captain Gregory Field 1674–78
Major John Blackmore 1678–90
Captain Joshua Johnson 1690–93
Captain Richard Kelling 1693–97
Captain Stephen Poirier 1697–1707
Captain Thomas Goodwin 1707–08
Captain John Roberts 1708–11
Captain Benjamin Boucher 1711–14
Captain Isaac Pyke 1714–19
Edward Johnson 1719–23
Captain John Smith 1723–27
Edward Byfield 1727–31
Captain Isaac Pyke 1731–38
John Goodwin 1738–40
Captain Robert Jenkins 1741–42
Major Thomas Lambert 1742
George Powell 1741–43
Colonel David Dunbar 1744–47
Captain Charles Hutchinson 1747–64
John Skottowe 1764–82
Daniel Corneille 1782–87
Colonel Robert Brooke 1787–1802
Colonel Robert Patton 1802–08
Alexander Beatson 1808–13
Colonel Mark Wilks 1813–16

Appendix: Governors of St Helena

Major-General Sir Hudson Lowe 1816–23
John Pine Coffin 1821–23
Brigadier-General Alexander Walker 1823–28
Brigadier-General Charles Dallas 1828–36

British Crown

Major-General George Middlemore 1836–42
Colonel Hamelin Trelawney 1842–46
Major-General Sir Patrick Ross 1846–50
Colonel Sir Thomas Gore Browne 1851–56
Sir Edward Hay Drummond Hay 1856–63
Admiral Sir Charles Elliot 1863–70
Vice-Admiral Charles Patey 1870–73
Hudson Janisch 1873–84
Lieutenant-Colonel Grant Blunt 1884–87
William Grey-Wilson 1887–97
Robert Sterndale 1897–1903
Lieutenant-Colonel Henry Gallwey 1903–12
Major Sir Harry Cordeaux 1912–20
Colonel Sir Robert Peel 1920–24
Sir Charles Harper 1925–32
Sir Steuart Spencer Davis 1932–38
Sir Guy Pilling 1938–41
Major William Gray 1941–47
Sir George Joy 1947–54
Sir James Harford 1954–58
Sir Robert Alford 1958–62
Sir John Field 1962–68
Sir Dermod Murphy 1968–71
Sir Thomas Oates 1971–76
Geoffrey Guy 1976–81
Sir John Massingham 1981–84
Francis Baker 1984–88
Robert Stimson 1988–91
Alan Hoole 1991–95
David Smallman 1995–99
David Hollamby 1999–2004
Michael Clancy 2004
Martin Hallam 2004–07
Andrew Gurr 2007–11
Mark Capes 2011–16
Lisa Phillips/Honan 2016–19
Dr Philip Rushbrook 2019–22
Nigel Phillips 2022–

Bibliography

Airs, Malcolm, 1998. 'The strange history of paper roofs', *Ancient Monuments Society Transactions* 42, pp. 36–62.
Amaral, Melchior Estacio do, 1604. *Tratado das batalhas e successos do Galeão Sanctiago com os Olandeses na ilha de Sancta Elena: e da náo Chagos com os Vngleses antre as Ilhas dos Açores* (Lisbon, Alvarez).
Annesley, George, Viscount Valentia, 1809. *Voyages and Travels to India, Ceylon, the Red Sea, Abyssinia, and Egypt in the Years 1802, 1803, 1804, 1805, and 1806*, vol. I (London, William Miller).
Ashmole, Myrtle, 2005. 'Endemic invertebrates, the airport and the St Helena Environment Charter', *Wirebird* 30, pp. 24–28.
Ashmole, Philip and Ashmole, Myrtle, 2000. *St Helena and Ascension Island: A Natural History* (Oswestry, Anthony Nelson).
Banks, Sir Joseph, 1958. *The Banks Letters. A Calendar of the Manuscript Correspondence of Sir Joseph Banks*, ed. W. R. Dawson (London, British Museum (Natural History)).
—— 1962. *The* Endeavour *Journal of Joseph Banks 1768–1771*, vol. II, ed. J. C. Beaglehole (Sydney, Public Library of New South Wales with Angus & Robertson).
Barlow, A. R., 1989. 'Forestry development on the island of St. Helena', *Commonwealth Forestry Review* 68, pp. 57–68.
Beatson, Alexander, 1816. *Tracts relative to the Island of St Helena … illustrated with views engraved by Mr William Daniell, from the Drawings of Samuel Davis* (London, G. & W. Nicol).
Bennett, Jim, 2014. '"The Rev. Mr Nevil Maskelyne FRS and myself": the story of Robert Waddington', in *Maskelyne, Astronomer Royal*, ed. Rebekah Higgitt (London, Royal Museums Greenwich and Robert Hale), pp. 59–88.
Bennett, Michael D., 2021. 'Slavery in early St Helena, part one: the "black servant" system', *Wirebird* 50, pp. 46–60.
Birch, Walter de Gray, 1880. *The Commentaries of the Great Afonso Dalboquerque*, vol. III, Hakluyt Society LXII (London, Hakluyt Society).
Biswas, Asit K., 1970. 'Edmond Halley, FRS, hydrologist extraordinary', *Notes and Records of the Royal Society* 25 no. 1, pp. 47–57.
Brooke, T. H., 1808. *A History of the Island of St Helena, from its discovery by the Portuguese to the year 1806* (London, Black, Parry & Kingsbury).

Bibliography

[Brooke, T. H.], 1810. *Papers Relating to the Devastation Committed by Goats on the Island of St. Helena ... Comprising Experiments, Observations & Hints connected with Agricultural Improvement and Planting* (St Helena, S. Solomon).

Bruce, Ian, 2015. 'St Helena Day', *Wirebird* 44, pp. 324–46; updated at http://sainthelenaisland.info/sthelenadayarticleianbruce.pdf.

Bryant, John, 2016. *RMS St Helena: Royal Mail Ship Extraordinary* (Ramsey, Ferry Publications).

Calendar of the Court Minutes etc. of the East India Company 1655–1659, ed. E. B. Sainsbury, intro. and notes by W. Foster, 1916 (Oxford, Clarendon Press).

Callander, John, 1768. *Terra Australis Cognita: or, Voyages to the Terra Australis, or Southern Hemisphere, during the Sixteenth, Seventeenth, and Eighteenth Centuries*, vol. III (Edinburgh, A. Donaldson).

Callaway, Ewen, 2016. 'Freedom in exile', *Nature* 540, pp. 184–87.

Cartwright, D. E., 1971. 'Tidal waves in the vicinity of Saint Helena', *Philosophical Transactions of the Royal Society. Series A. Mathematical and Physical Sciences* 270 no. 1210, pp. 603–46.

Castell, Robin, 2011. *William Burchell (1781–1863) St Helena (1805–1810)* (St Helena, Castell Collection).

Cawood, John, 1979. 'The Magnetic Crusade: science and politics in early Victorian Britain', *Isis* 70 no. 4, pp. 492–518.

Chalmers, J. H., 1869. Report on the Experiment of Establishing the Chinchona Plant in St Helena: from 7th July 1868, to 17th December 1869 (n.p.).

Chancellor, Gordon Russell, 1990. 'Charles Darwin's St Helena model notebook', *Bulletin of the British Museum (Natural History)*, hist. ser. 18 no. 2, pp. 203–28.

Chancellor, Gordon and Van Wyhe, John, 2009. *Charles Darwin's Notebooks from the Voyage of the Beagle* (Cambridge, Cambridge University Press).

Chapman, Allan, 1994. 'Edmond Halley's use of historical evidence in the advancement of science', *Notes and Records of the Royal Society of London* 48 no. 2, pp. 167–91.

Chartrand, René, 2011. 'St. Helena local militia, c. 1837–1840', *Journal of the Society for Army Historical Research* 89, pp. 261–63.

Clements, Bill, 2006. 'Second World War defences on St Helena', *Wirebird* 3, pp. 11–15.

—— 2007. 'St Helena: South Atlantic fortress', *Fort* 35, pp. 75–90.

Cleverly, Les, 1989. *W. J. Burchell. A Short Biography* (privately printed).

Collection of Statutes, 1786. *A Collection of Statutes concerning the Incorporation, Trade, and Commerce of the East India Company, and the Government of the British Possessions in India* (London, Eyre & Strahan).

Cook, Alan, 1998. *Edmond Halley. Charting the Heavens and the Seas* (Oxford, Clarendon Press).

Cronk, Q. C. B., 1987. 'The history of endemic flora of St Helena: a relictual series', *New Phytologist* 105, pp. 509–20.

―― 1988. 'W. J. Burchell and the botany of St Helena', *Archives of Natural History* 15, pp. 45–60.
―― 1989. 'The past and present vegetation of St Helena', *Journal of Biogeography* 16, pp. 47–64.
―― 2000. *The Endemic Flora of St Helena* (Oswestry).
Daly, Reginald A., 1927. 'The geology of Saint Helena Island', *Proceedings of the American Academy of Arts and Sciences* 62 no. 2, pp. 31–92.
Dampier, William, 1937. *A New Voyage Round the World*, intro. Sir Albert Gray (London, A. & C. Black).
Darwin, Charles, 1859. *On the Origin of Species by means of Natural Selection* (London, John Murray).
―― 1860. *Journal of Researches into the Natural History and Geology of the Countries visited during the Voyage of HMS Beagle round the World* (London: John Murray).
―― 1890. *A Naturalist's Voyage. Journal of Researches into the Natural History and Geology of the Countries visited during the Voyage of HMS Beagle round the World*, new edn (London: John Murray).
―― 1933. *Charles Darwin's Diary of the Voyage of H.M.S. 'Beagle'*, ed. N. Barlow (Cambridge, Cambridge University Press).
David, Andrew, 1993. 'Bligh's successful breadfruit voyage', *Royal Society of Arts Journal* 141, pp. 821–24.
Denholm, Ken, 2006. *South Atlantic Fortress* (St Helena, National Trust).
―― n.d. *From Signal Gun to Satellite. A History of Communications on the Island of St Helena* ([St Helena]).
Desmond, Ray, 1992. *The European Discovery of the Indian Flora* (London, Royal Botanic Gardens, Kew).
―― 1999. *Sir Joseph Dalton Hooker, Traveller and Plant Collector* (Woodbridge, Antique Collectors' Club and RBG Kew).
[Duncan, Francis], 1805. *A Description of the Island of St. Helena; containing observations on its singular structure and formation; and an account of its climate, natural history, and inhabitants* (London, R. Phillips).
Ekoko, A. E., 1984. 'The theory and practice of imperial garrisons: the British experiment in the South Atlantic, 1881–1914', *Journal of the Historical Society of Nigeria* 12 nos. 1–2, pp. 133–48.
Fenton, Edward, 1957. *The Troublesome Voyage of Captain Edward Fenton, 1582–1583*, ed. E. G. R. Taylor, Hakluyt Society, 2nd ser. CXIII (London, Hakluyt Society).
Field, Margaret, 1998. *The History of Plantation House and Grounds, St Helena, 1673–1967* (Penzance, Patten Press).
Fogg, G. E., 1992. *A History of Antarctic Science* (Cambridge, Cambridge University Press).
Forster, George, 2000. *A Voyage Round the World*, ed. Nicholas Thomas and Oliver Berghof, with Jennifer Newell (Honolulu, University of Hawai'i Press).

Bibliography

Forster, Johann Reinhold, 1982. *The Resolution Journal of Johann Reinhold Forster 1772–1775*, ed. M. L. Hoare, Hakluyt Society, 2nd ser. no. CLII (London, Hakluyt Society).

Forsyth, William, 1853. *History of the Captivity of Napoleon at St Helena: from the letters and journals of Lieut.-Gen. Sir Hudson Lowe*, 3 vols (London, John Murray).

Foster, Sir William, 1919. 'The acquisition of St Helena', *English Historical Review* 36 no. 135, pp. 281–89.

Fox, Colin, 2017. *A Bitter Draught. St Helena: The Abolition of Slavery 1792–1840* (Elvedon, Norfolk, Friends of St Helena).

—— 2021. 'Napoleon's coffin', *Wirebird* 50, pp. 5–24.

Frankl, P. J. L., 2006. 'The exile of Sayyid Khalid bin Barghash Al-BuSa'idi', *British Journal of Middle Eastern Studies* 33 no. 2, pp. 161–77.

Giles, Frank, 2001. *Napoleon Bonaparte: England's Prisoner* (New York, Carroll & Graf).

Gill, Mrs [Isobel S. B.], 1878. *Six Months in Ascension. An unscientific account of a scientific expedition* (London, John Murray).

Godwin, Francis, 1638. *The Man in the Moone: or a discourse of a voyage thither*, 2nd edn (London, Joshua Kirton).

Gosse, Philip, 1990. *St Helena, 1502–1938* (Oswestry, Anthony Nelson; first published London, Cassell, 1938).

Gray, Steven, 2018. *Steam Power and Sea Power. Coal, the Royal Navy, and the British Empire, c.1870–1914* (London, Palgrave Macmillan).

Griggs, William, 1909. *Relics of the Honourable East India Company* (London, Bernard Quaritch).

Grove, A. T., 2015. 'St Helena as a microcosm of the East India Company world', in *The East India Company and the Natural World*, ed. V. Damodaran, A. Winterbottom and A. Lester (London, Palgrave Macmillan), pp. 249–69.

Grove, Richard, 1993. 'Conserving Eden: the (European) East India Companies and their environmental policies on St Helena, Mauritius and in Western India, 1660 to 1854', *Comparative Studies in Society and History* 35 no. 2, pp. 318–51.

—— 1995. *Green Imperialism. Colonial Expansion, Tropical Island Edens and the Origins of Environmentalism, 1600–1860* (Cambridge, Cambridge University Press).

Gualtieri, Guido, 1586. *Relationi della venuta degli Ambasciatori Giaponesi a Roma fino alla partita di Lisbona* (Rome, Francesco Zannetti).

Hakluyt, Richard, 1903–05. *The Principal Navigations Voyages Traffiques & Discoveries of the English Nation* (Glasgow, James Maclehose & Sons).

Haliburton, R. G., 1890. *Letters on the Withdrawal of the Garrison from 'The Citadel of the South Atlantic', St Helena* (Jamestown, B. Grant).

Halley, Edmond, 1679. *Catalogus Stellarum Australium sive Supplementum Catalogi Tychonici* (London, Thomas James).

―― 1686. 'An historical account of the trade winds, and monsoons, observable in the seas between and near the tropicks, with an attempt to assign the phisical cause of the said winds', *Philosophical Transactions* 16, pp. 153–68.

―― 1932. *Correspondence and Papers of Edmond Halley*, ed. E. F. MacPike (Oxford, Oxford University Press).

Hamilton, Douglas and McAleer, John (eds), 2021. *Islands and the British Empire in the Age of Sail* (Oxford, Oxford University Press).

Hearl, Trevor, 1999. 'St Helena's 16th century astronaut. The adventures of Domingo Gonsales, the speedy messenger', *Wirebird* 42, pp. 25–29.

Hearl, Trevor W., 2013. *St Helena Britannica. Studies in South Atlantic Island History*, ed. A. H. Schulenburg (London, Society of Friends of St Helena).

Herbert, Sir Thomas, 1677. *Some Years Travels into Divers Parts of Africa, and Asia the Great* (London, R. Everingham for R. Scott *et al.*).

Heunis, Victoria Regina, 2019. 'Anglo-Boeroorlog Boerekrygsgevangenekuns, 1899–1902', PhD dissertation, University of Pretoria.

Hinks, Arthur R., 1944. *Maps and Survey*, 5th edn (Cambridge, Cambridge University Press).

Howarth, Richard J., 2007. 'Gravity surveying in early geophysics, 1: from timekeeping to figure of the Earth', *Earth Sciences History* 26 no. 2, pp. 201–28.

Howse, Derek, 2004. 'Maskelyne, Nevil (1732–1811)', *Oxford Dictionary of National Biography* (Oxford, Oxford University Press).

Janisch, Hudson R., 1885. *Extracts from the St Helena Records* (St Helena, The Guardian).

Johnson, Manuel J., 1835. *A Catalogue of 606 Principal Fixed Stars in the Southern Hemisphere, deduced from observations made at the Observatory, St Helena* (London, Royal Astronomical Society/Honourable East India Company).

Jones, Sir Harold Spencer, 1957. 'Halley as an astronomer', *Notes and Records of the Royal Society* 12 no. 2, pp. 175–92.

Joyce, Miriam, 2000. 'The Bahraini Three on St Helena, 1956–1961', *Middle East Journal* 4, pp. 613–23.

Kinns, Roger, 2021. 'Time signals for mariners in the Atlantic islands and West Africa', *Journal of Astronomical History and Heritage* 24 no. 2, pp. 315–36.

Kitching, Christopher, 1937. *A Handbook and Gazetteer of the Island of St Helena Including a Short History of the Island under the Crown 1834–1902* (St Helena, printed privately).

―― 1947. 'The St Helena regiments of the East India Company', *Journal of the Society for Army Historical Research* 25, pp. 2–8.

―― 1950. 'The loss and recapture of St Helena, 1673', *Mariner's Mirror* 36, pp. 58–68.

Lambdon, P., Darlow, A., Clubbe, C. and Cope, T., 2013. '*Eragrostis episcopulus* – a newly described grass species endemic to the island of St Helena …', *Kew Bulletin* 68, pp. 121–31.

Lancaster, Sir James, 1940. *The Voyages of Sir James Lancaster to Brazil and the East Indies 1591–1603*, ed. Sir William Foster (London, Hakluyt Society).

Bibliography

Leguat, François, 1720. *Voyages et aventures de François Leguat, & de ses compagnons en deux isles désertes des Indes Orientales. Avec la relation des choses les plus remarquables qu'ils ont observées dans l'Isle Maurice ... dans l'Isle St. Hélene, & en autres endroits de leur route* (London, David Mortier).

Levy, Martin, 1998. 'Napoleon in exile: the houses and furniture supplied by the British government for the emperor and his entourage on St Helena', *Furniture History* 34, pp. 1–211.

Van Linschoten, Jan Huygen, 1598. *John Huighen van Linschoten, his Discours of Voyages into ye Easte & West Indies: devided into Foure Bookes* (London, John Wolfe).

Van Linschoten, J. H., 1885. *The Voyage of John Huyghen van Linschoten to the East Indies*, vol. II, ed. P. A. Tiele, Hakluyt Society LXXI (London, Hakluyt Society).

Ly-Tio-Fane, Madeleine, 1996. 'Botanic gardens: connecting links in plant transfer between the Indo-Pacific and Caribbean regions', *Harvard Papers in Botany* 1 no. 8, 7–14.

MacGregor, Arthur, 1992. 'Deer on the move: relocation of stock between game parks in the sixteenth and seventeenth centuries', *Anthropozoologica* 16, pp. 167–79.

—— 2012. *Animal Encounters. Human and Animal Interaction in Britain from the Norman Conquest to World War I* (London, Reaktion Books).

—— 2018. 'Fortifications in the sand: modelling and military education in the nineteenth century', *Journal of the Society for Army Historical Research* 96 no. 387, pp. 251–66.

—— 2023. *The India Museum Revisited* (London, V&A and UCL Press).

Maitland, Rear-Admiral Sir Frederick Lewis, 1904. *The Surrender of Napoleon*, new edn, ed. W. K. Dickson (Edinburgh and London, Blackwood).

Major, Andrea, 2012. *Slavery, Abolitionism and Empire in India, 1772–1843* (Liverpool, Liverpool University Press).

Malin, S. R. C. and Barraclough, D. R., 1991. 'Humboldt and the Earth's magnetic field', *Quarterly Journal of the Royal Astronomical Society* 32 no. 3, pp. 279–93.

Martineau, Gilbert, 1968. *Napoleon's St Helena* (London, John Murray).

—— 1971. *Napoleon Surrenders*, trans. F. Partridge (London, John Murray).

Maskelyne, Nevil, 1761. 'Account of the observations made on the Transit of Venus June 6, 1761, in the island of St Helena', *Philosophical Transactions* 52, pp. 196–201.

—— 1762. 'Observations on the tides in the island of St. Helena', *Philosophical Transactions* 99, pp. 586–91.

—— 1762. 'Observations of the tides made in the harbour at James's Fort, St. Helena', *Philosophical Transactions* 99, pp. 592–606.

Mason, Charles, 1761. 'Observations for proving the going of Mr. Ellicott's clock, at St. Helena', *Philosophical Transactions* 52, pp. 534–39.

McCracken, Donal P., 1997. *Gardens of Empire. Botanical Institutions of the Victorian British Empire* (Leicester).

―― 2022. *Napoleon's Garden Island. Lost and Old Gardens of St Helena, South Atlantic Ocean* (London, Kew Publishing, Royal Botanic Gardens).

McCulloch, M. Neil, 1991. 'Status, habitat and conservation of the St Helena Wirebird *Charadrius sanctaehelenae*', *Bird Conservation International* 1, pp. 361–92.

McKay, H. M., 1934. 'William John Burchell in St Helena, 1805–1810', *South African Journal of Science* 31, pp. 481–89.

Mellis, John Charles, 1870. 'Notes on the birds of the island of St Helena', *The Ibis* new ser. 6, pp. 97–107.

―― 1875. *St Helena: A physical, historical, and topographical description of the island, including its geology, fauna, flora, and meteorology* (London, L. Reeve & Co.).

Mundy, Peter, 1914. *The Travels of Peter Mundy, in Europe and Asia, 1608–1667*, vol. II: *Travels in Asia 1628–1634*, Hakluyt Society, 2nd ser. no. 35 (London, Hakluyt Society).

Murray, Paul, 2003. '"On Saint Helena's bleak shore": Free State plans to intern Republican prisoners', *History Ireland* (Spring), pp. 10–11.

Napoleon's Appeal, 1817. *Napoleon's Appeal to the British Nation, on his Treatment at Saint Helena. The Official Memoir, dictated by him, and delivered to Sir Hudson Lowe* (London, William Hone).

Van Niekerk, J. P., 2009. 'The role of the Vice-Admiralty Court at St Helena in the abolition of the trans-Atlantic slave trade: a preliminary investigation', *Fundamina, A Journal of Legal History* 15 no. 1, pp. 69–111; no. 2, pp. 1–56.

Nowak-Kemp, Malgosia, 2018. 'William Burchell in Southern Africa, 1811–1815', in *Naturalists in the Field. Collecting, Recording and Preserving the Natural World from the Fifteenth to the Twenty-First Century*, ed. A. MacGregor (Leiden and Boston, Brill), pp. 500–49.

Ólafsson, Jón, 1932. *The Life of the Icelander Jón Ólafsson*, trans. B. Phillpotts, ed. R. Temple and L. M. Anstey, Hakluyt Society, 2nd ser. no. 68 (London, Hakluyt Society).

Oldenburg, Henry, 1986. *The Correspondence of Henry Oldenburg*, ed. and trans. A. R. Hall and M. B. Hall (London and Philadelphia, Taylor & Francis).

Olson, Storrs L., 1975. 'Paleornithology of St Helena Island, South Atlantic Ocean', *Smithsonian Contributions to Paleobiology* 23, pp. 1–43.

O'Meara, Barry E., 1822. *Napoleon in Exile; or, A Voice from St Helena*, 2 vols (London, Simpkin & Marshall).

Osorio, Jerome, 1752. *The History of the Portuguese, during the Reign of Emmanuel*, ed and trans. J. Gibbs (London, A. Millar).

Palmer, Edmund, 1858–59. 'Notes on the island of St Helena; to accompany his new map of that island', *Proceedings of the Royal Geographical Society* 3 no. 6, pp. 363–64.

Pearson, Andrew, 2016. 'Waterwitch: a warship, its voyage and its crew in the era of anti-slavery', *Atlantic Studies* 13, pp. 99–124.

Bibliography

Pearson, Andrew, Jeffs, B., Witkin, A. and MacQuarrie, H., 2011. *Infernal Traffic: Excavation of a Liberated African Graveyard in Rupert's Valley, St Helena* (York, Council for British Archaeology).

Pillans, T. Dundas, 1913. *The Real Martyr of St Helena* (New York, McBride, Nast & Co.).

Van der Pijl-Ketel, C. L. (ed.), 1982. *The Ceramic Load of the Witte Leeuw (1613)* (Amsterdam, Rijksmuseum).

Powell, Dulcie, 1977. 'The voyage of the plant nursery, HMS *Providence*, 1791–1793', *Economic Botany* 31 no. 4, pp. 387–431.

Prior, James, 1819. *Voyage along the Eastern Coast of Africa to Mosambique, Johanna and Quiloa; to St Helena* (London, printed for Sir Richard Phillips).

Pyrard, François, 1890. *The Voyage of François Pyrard, of Laval, to the East Indies, the Maldives, the Moluccas and Brazil*, Hakluyt Society LXXX (London, Hakluyt Society).

Report of the St Helena Industrial Exhibition, 1874. Report of the St Helena Industrial Exhibition, for 1874. Presented to His Excellency the Governor (St Helena).

Roberts, Andrew, 2015. *Napoleon the Great* (London, Penguin Books).

Robinson, Tim, 2008. *William Roxburgh, the Founding Father of Indian Botany* (Chichester, Phillimore and RBG Edinburgh).

Rogers, Francis, 1936. 'The diary of Francis Rogers', in *Three Sea Journals of Stuart Times*, ed. B. S. Ingram (London, Constable & Co.).

Royle, Stephen, 1997. '"Alexander the rat". F. W. Alexander, chief censor, Deadwood Camp, St Helena', *Wirebird* 15, pp. 17–21.

—— 1998. 'St Helena as a Boer prisoner of war camp, 1900–2: information from the Alice Stopford Green papers', *Journal of Historical Geography* 24 no. 1, pp. 53–68.

—— 2007. *The Company's Island. St Helena, Company Colonies and the Colonial Endeavour* (London, I.B. Tauris).

—— 2019. 'Island history, not the story of islands: the case of St Helena', *Shima* 13 no. 1, pp. 44–55.

Sainsbury, E. (ed.), 1912. *A Calendar of the Court Minutes etc. of the East India Company* (Oxford, Clarendon Press).

St Helena Government, n.d. *St Helena Airport and Supporting Infrastructure. Planning Statement*, https://www.sainthelena.gov.sh/wp-content/uploads/2017/08/Planning-Statement.pdf.

St Helena Industries Ltd., [1895]. *Notes on the Fisheries and other Industries capable of being undertaken in the Island of St Helena* (St Helena).

Sandoval-Velasco, M. *et al.*, 'The genetic origins of Saint Helena's liberated Africans', *BioRxiv preprint*, https://doi.org/10.1101/787515.

Shine, Ian, 1970. *Serendipity in St Helena. A Genetical and Medical Study of an Isolated Community* (Oxford, Pergamon Press).

Stanwood, Owen, 2020. *The Global Refuge. Huguenots in an Age of Empire* (Oxford, Oxford University Press).

Bibliography

Stern, Philip J., 2007. 'Politics and ideology in the early East India Company-State: the case of St Helena, 1673–1709', *Journal of Imperial and Commonwealth History* 35 no. 1, pp. 1–23.

Tatham, W. G. and Harwood, K. A., 1974. 'Astronomers and other scientists on St Helena', *Annals of Science* 31 no. 6, pp. 489–510.

Tavernier, Jean-Baptiste, 1678. *The Six Voyages of John Baptista Tavernier, a Noble Man of France now living, through Turky into Persia, and the East Indies, finished in the year 1670, made English by J. P.* (London, R., L. and M. Pitt).

Thrower, Norman J. W., 1981. *The Three Voyages of Edmond Halley in the Paramore 1698–1701*, Hakluyt Society, 2nd ser. 156–57 (London, Hakluyt Society).

Todd, David P., 1890. 'The United States Scientific Expedition to West Africa, 1899', *Nature* 42 no. 1070, pp. 8–10.

—— 1891. 'Additional results of the United States Scientific Expedition to West Africa', *Nature* 43 no. 1120, pp. 563–65.

Turner, I. M., 2016. 'Notes relating to William Roxburgh's study of the flora of St Helena', *Kew Bulletin* 71 no. 2, pp. 1–5.

Turrill, W. B., 1948. 'On the flora of St Helena', *Kew Bulletin* 3 no. 3, pp. 358–62.

Vassie, J. M., Woodworth, P. L. and Holt, M. W., 2004. 'An example of North Atlantic deep-ocean swell impacting Ascension and St Helena Islands in the central South Atlantic', *Journal of Atmospheric and Oceanic Technology* 21, pp. 1095–103.

Van de Velde, Stéphane, 2011. 'Chasing the slavers: the establishment of the Vice Admiralty Court of St Helena and its early cases', *Wirebird* 40, pp. 3–14.

Warner, Brian, 1982. 'Manuel Johnson and the St Helena Observatory', *Vistas in Astronomy* 25, pp. 383–409.

Warner, Oliver, 1953. *Captain Marryat. A Rediscovery* (London, Constable & Co.).

Webster, W. H. B., 1834. *Narrative of a Voyage to the Southern Atlantic Ocean in the Years 1828, 29, 30: Performed in H. M. Sloop Chanticleer, under the command of the Late Captain Henry Foster, F. R. S. &c. by order of the Lords Commissioners of the Admiralty* (London, Richard Bentley).

Wills, Mary, 2019. *Envoys of Abolition. British Naval Officers and the Campaign against the Slave Trade in West Africa* (Liverpool, Liverpool University Press).

Winterbottom, Anna E., 2010. 'Company culture: information, scholarship, and the East India Company settlements 1660–1720s', PhD thesis, University of London.

Woolf, Harry, 1956. 'British preparations for observing the Transit of Venus of 1761', *William and Mary Quarterly* 13, pp. 499–518.

Yu, Po-ching, 2015. 'Chinese seamen in London and St Helena in the early nineteenth century', in *Law, Labour and Empire. Comparative Perspectives on Seafarers, c.1500–1800*, ed. M. Fusaro, B. Allaire, R. J. Blakemore and T. Vanneste (Basingstoke, Palgrave Macmillan), pp. 287–303.

Index

Abyssinia, Patriarch of   13, 34
acclimatisation, of plants   133, 139
   of troops   79
Addiscombe Military Seminary   80, 146n.97
Africa   5, 31, 56, 79, 147n.98, 137, 147, 167, 170, 185, 194
Agricultural and Horticultural Society   50, 53
agricultural improvement   45, 49–50, 53, 61
Agriculture and Forestry Department   203
airport   199–202, Plate 29
alarm guns   72, 83–4, 79n.41, 84
Alexander, F. W.   177n.24
Alford, Sir Robert   187, 208
Americas   5, 12, 36, 50, 56, 147, 150, 167, 191–2
Anderson, James   50
Anglo-Zulu War   172
Angola   3, 168n.10, 170
animals   6, 18, 24–5, 49, 53, 58, 145–6,
   mammal species
      asses   60, 129
      cats   22, 25–6, 44
      cattle, oxen   6, 25, 32–4, 49n.62, 52, 57, 59n.92, 60, 129, 181
      deer   10, 25, 60
      dogs   22, 25–6, 155n.13
      dolphins   28
      donkeys   129n61, 181
      goats   9–10, 17, 22, 24–5, 33, 44, 57–9, Fig. 4
      grampuses   28
      horses   25, 48, 49n.62, 129, 158n.20
      manatees   28
      mice   129
      mules   129n.61, 159n.23, 181

   pigs/hogs/swine   9, 20n.8, 22, 25, 34, 44, 56, 59–60
   rabbits   26
   rats   22, 26, 33, 45, 129, 155
   sea cows   28
   'sea lioness'   28, Fig. 5
   seals   28
   sheep   33–4, 57, 59–60
   whales   28
   reptile species
      turtles   29
Antarctica   1, 54, 120, 122, 146
Antommarchi, Francesco   160
arak, consumption of   17, 45n.53, 47, 52
Armstrong, George   117
Arnott, Archibald   51
artillery   71–2, 74–7, 162
   6-inch breech-loading guns   81, 82n.55
   12-pounder guns   77
   cannon   15, 65, 71, 74, 76
   carronades   74
   Elswick coastal defence guns   82, Plate 14
   howitzers   74
   mortars   74
Ascension Island   3, 27, 119n.32, 125, 152, 188, 190n.1, 191, 195, 197, 199n.11, 204
Ashmole, Philip and Myrtle   148
astronomical instruments   108–19, Fig. 15
astronomical observatories   109, 115–19, Figs 16–17
Atlantic Ocean   1, 193, 204
   Benguela Current   3
   Mid-Atlantic Ridge   1
   South Atlantic Gyre   3
Australia   126, 145, 193–4

Bahraini Three   186–8
Baker, Francis   208
Balcombe, Betsy   153, 158n.20

Index

Balcombe, William   153, 159n.26, 165n.37
Banks, Sir Joseph   37, 44, 45n.54, 47–9, 60, 62, 114, 126–8, 130, 132–3, 139, 141, 205
Barbados   36
Barker, Edmund   13
Bass, Captain William   71
Bathurst, Earl   159n25
Beale, Captain Anthony   70, 72–3, 207
Beatson, Colonel Alexander   49–50, 56, 58, 59n.92, 61, 67n.10, 83, 133, 140–1, 144, 207
Belfast, George Hamilton Chichester, Lord   167n.7
Belgrave, Sir Charles   186
Bellasis, George Hutchins   Plates 1, 15
Bendall, Captain   37
Benjamin, Charles and George   4, 149
Berry, Dr Andrew   141n.81
Bertrand, Count and Countess   152, 154, 159–60
Best, Thomas   20
Biosecurity St Helena   202
birds   4, 65, 18, 21–2, 26, 34, 129, 159n.24, Fig. 4
   crakes   5
   cuckoos   5
   ducks   9, 44
   fowls/hens   9, 21, 34, 44, 60
   frigatebirds   5
   geese   44
   Glossy ibis (*Plegadis falcinellus*)   22n.18
   guineafowl   21, 129
   hoopoes   5
   mynahs   60n.100
   partridges   21, 26, 129
   petrels   5
   pheasants   21, 129
   pigeon/doves (*Dysmoropelia dekarchikos*)   21
   plovers   5-6, 22
   puffins   5
   rails   5
   rice birds/paddies (*Loxia oryzivora*)   21n.15
   shearwaters   5
   tropicbirds, red-billed   5
   turkeys   21, 44
   wirebirds (*Charadrius sanctaehelenae*)   5, 22, 199, Plate 9
Black Oliver   35n.27
Blackmore, Major John   207
Bligh, Captain William   33, 51, 128, 130-3
Blunt, Lieutenant-Colonel Grant   208
Board of Longitude   115
boats, fishing boats   27–8, 37, 56
Boer War prisoners   82, 174–82, Figs. 10, 28–31, 190, Plates 25–6
Bosanquet, Jacob   205
Boucher, Captain Benjamin   60, 207
Bowen, Captain Robert   21, 31, 35
Bradley, James   113
Brahe, Tycho   108, 112n.12
Brazil   3, 35n.27, 37n.35, 63, 68, 70, 125, 139n.78, 142, 168, 191
brick, as building material   32, 45
British Guiana   172
Brooke, Colonel Robert   40, 49n62, 55, 77, 79, 83, 131, 141, 207
Brooke, T. H.   8, 141
Bry, Theodore de   1, Figs 1–2
Bunbury, Major-General Sir Henry   151
Burchell, William   26, 84, 133, 136–7, 139–43, 146n.96, Fig. 20a, Plate 19
Burton, Francis   160
Byfield, Edward   44, 58n.90, 59, 62, 74, 207

cabinetmaker's work   64
Cable & Wireless   192, Fig. 33
Caesar, Samuel   Fig. 6
Campbell, General Sir Archibald   77
Canary Islands   192, 195
Candolle, A. P. de   142n.86
cannery   56
Canton   43
Cape of Good Hope   3, 8, 12, 49n.62, 53, 62–3, 65, 68n.15, 79, 113n.14, 114–15, 117, 121n.39, 122, 126–9, 131, 139, 141–3, 145–6, 152, 159n.25, 172, 181n33, 193–4
Cape Town   115, 119n.32, 191, 195, 197, 199
Cape Verde Islands   21, 31, 191
Capes, Mark   208
carts   48–9
Cason, Captain   77
Castle Mail Packets Company   194

Index

Cavendish, Captain Thomas   12, 19, 21–2, 34–5
Ceylon   40, 147, 174
Chalmers, J. H.   53
Charles I, King   32n6
Charles II, King   108, 112n.12
China   11, 122, 139, 141, 191
Chinese presence on St Helena   43, 159
Cipriani, [Franceschi]   151n.4
Clancy, Michael   208
clays (coloured)   141
Clerk, James   108n.3
coal, coaling facilities   81, 137, 194
Cockburn, Rear-Admiral Sir George   152–3, 155, 157
Colonial and Foreign Fibre Co.   54
Colonial Prisoners Removal Act (1869)   187
Coney, Captain Richard   207
Cook, Captain James   38, 49, 114, 126, 128
Cordeaux, Major Sir Harry   183, 208
Corneille, Daniel   59, 128, 207
Cornwallis, Charles, Marquess   77
Covid-19   198, 202
Cox, Mr   53
Cromwell, Oliver   30, 66
Cromwell, Richard   66
Cronjé, General Piet   175, 177, 180n29, 181n.33
Cronk, Quentin   4n.9, 142, 148–9
Cuba   168

Dallas, Brigadier-General Charles   194, 208
Dampier, William   20, 47, 52, 72
Dancoisne-Martineau, Michel   151n.5, 153n.8
Daniel, Thomas and William   141
Dar-es-Salaam   183, 185
Darwin, Charles   18n.3, 41–2, 76, 79n.41, 80n.45, 144–6, 164, 166, 200, Figs 20c, 35
Davis, Sir Steuart Spencer   208
Dawson, Mr   140
Deason, Thomas   45, Fig. 9
Decrès, Admiral Denis   79
Desduguieres, M.   73n.27
Dinuzulu   166, 172–4, Fig. 27
Diplomatic Wireless Service   192

Dixon, Captain Charles   84
Dixon, Jeremiah   113
Domaines Nationaux à Sainte-Hélène   165n.37
Drummond, Captain Charles   139
Drummond Hay, Sir Edward   63, 208
Dunbar, Colonel David   45, 49n.66, 207
Duncan, Mr   139
Dutch East India Company (VOC)   16, 68, 70
Dutch presence on St Helena   14–16, 19, 25, 27, 32, 35, 65, 68, 72, 76, 127, 166n.2, 177
  proclamation of ownership   15, Plate 8
Dutton, Captain John   20, 30, 66–8, 207, Fig. 11

East India Company (EIC)   16, 20, 22, 25, 30–3, 35, 37n35, 38, 40, 42–3, 47–8, 50–3, 55, 58–9, 63, 64n111, 65–81, 83, 108, 113, 117, 128, 130, 133, 137, 140–1, 143–4, 147, 153, 166n.1, 190, 193, 197, 205
  charters of (1657, 1658, 1661, 1673)   66, 190
Eastern Telegraph Company   181, 191
eggs, consumption of   22, 29, 34
Elliot, Admiral Sir Charles   54, 63, 208
Elliot, Captain   48
Emmett, Major Anthony   154

Falkland Islands   193n.6, 195, 199, 204
Fallows, Revd Fearon   115
fencing and enclosure   33, 57–8, 62
Fenton, Edward   11
Field, Captain Gregory   72, 207
Field, Sir John   208
firewood   33, 45, 63, 137
fish   14, 27, 34, 47n.57, 60, 64, 181, 203
  bulls' eyes   27
  cavalloes   27
  coal-fish   27
  flying fish   27
  jacks   27
  horse mackerel   27
  mackerel   27
  old-wives   27
  sharks   28
  soldiers   27

Index

fishing industry 55–6
fishing rods 64
flags, signalling 83–4
Flamsteed, John 108–9, 112
flax mills 54
Forster, George 21n.15, 38, 43, 49, 58, 62, 129
Forster, Johann Reinhold 128–30
Fort Cormantine 35
fortifications and gun emplacements 65–82, 116
  Martello towers 77
  musketry towers 77
Fortune, John 37n.35
Foster, Captain Henry 120–2
fowl pest 44
French presence on St Helena 14, 52, 70, 73, 79n.43, 164–5

Gabon 170
Galapagos Islands 145–6
Gallwey, Lieutenant-Colonel Henry 208
gardens 51, 60, 61, 127, 137, 139–41, 153, 167, Fig. 21
Gargen, Henry 38n.39
garrison troops 33, 35, 38, 40, 58, 62, 65–6, 68, 70, 72, 76n.36, 77, 79n.43, 81, 115, 136, 152, 158–9
  20th (East Devonshire) Regiment 152, 162
  21st Light Dragoons 162
  53rd (Shropshire) Regiment 152, 154
  66th (Berkshire) Regiment 82n.56, 152, 160, 162
  91st (Argyllshire) Regiment 80
  Gloucestershire Regiment / Royal North Gloucestershire Regiment 82n.56, 180–1
  Royal Artillery 122, 162
  St Helena Coast Battery 81–2
  Royal Engineers 75, 84–5, 154
  St Helena Artillery 72, 74, 80, 115, 117, 147, 152
  St Helena Infantry 72, 80
  St Helena Local Militia 72 , 77, 81, Fig. 14a
  St Helena Regiment 80, 116, 152, 162
  St Helena Rifles 82, Fig. 14b
  St Helena Volunteer Rifles (Sharpshooters) 82
  St Helena Volunteers 81, 162, 175
  West India Regiment 175, 180n30
  Wiltshire Regiment 82n.56
General Screw Steam Shipping Company 194
George V, King 183
German East Africa 183
Gill, Isobel and David 119
Godwin, Bishop Francis 6, Fig. 2
Goodwill Zwelithini, King 166
Goodwin John 207
Goodwin, Captain Thomas 207
Gore Brown, Colonel Sir Thomas 63, 208
Gourgaud, General Gaspard 152
Gray, Major William 208
Great Exhibition 51n.73
Green Alice Stopford 177n.25
Grey-Wilson, William 56, 208
Gurr, Andrew 208
guano 49n.65, 141
Guinea 35, 36n.30

Halkett, Captain 141
Hallam, Martin 208
Halley, Edmond 57n87, 108–14, 117, 119–20, 124, Plates 17–18
Harford, Sir James 208
Harper, Sir Charles 208
Herbert, Sir Thomas 9n.5, 14
Herschell, Sir John 117n.29
Hesse, Revd C.H.F. 139
Hodson, Judge Charles 167n.8
Hollamby, David 208
Holland, Lord 151n.6
honey 64
Honeywood, Richard 34
Hooker, Joseph Dalton 53, 146–8, Fig. 20d
Hoole, Alan 208
Humboldt, Alexander von 122
Hutchinson, Captain Charles 207

Imray, James Plate 2
India 8, 10, 13, 15, 26, 31, 36, 42–3, 60n.100, 79–80, 129–30, 139, 141n.82, 143, 147, 153n.7, 174, 191, 205, Plate 3
India Museum 45n.55

Index

Indian Ocean   3, 5, 8, 21
   Agulhas Current   5, 130
Indian presence on St Helena   43
Industrial Exhibition (Jamestown)   64
invertebrates   26–7, 199
   beetles   27
      She Cabbage beetle   6
      cochineal beetle   50, 141n.81
      giant ground beetle (*Aplothorax burchelli*)   26
   bees   203
   blushing snail (*Succinea sanctaehelenae*)   6
   caterpillars 129
   cockroaches 155
   giant earwig (*Labidura herculeana*) 26
   mosquitoes 155
   silkworms   51
   spiders (*Argyrodes mellissi*)   27, 147–8
   shore woodlouse (*Littorophiloscia*)   6
   spiky yellow woodlouse (*Pseudolaureola atlantica*)   26
   termites   62
Irish Republicans   185–6
irrigation schemes   44, 57n.87, 62

James, Duke of York (later James II)   67
Janisch, Hudson   208
Japanese presence on St Helena   12n16
Jenkins, Captain Robert   207
Johanna (Comoro Islands)   36
Johannesburg   202
Johnson, Edward   207
Johnson, Captain Joshua   207
Johnson, Manuel   115–17, Fig. 17
Joy, Sir George   208

Keigwin, Captain Richard   35n.27, 70, 207
Kelling, Captain Richard   207
Kendall, Abraham   13
Kenya   183
Kerr, Robert   139n77
Kew, Royal Gardens/Royal Botanic Gardens   51, 53–4, 128, 131–3, 136, 139, 142, 147
Knox, Captain   36
Kroomen   167
Kydd, Robert   143

lace making   64
Lambert, A. B.   142
Lambert, Major Thomas   207
Lancaster, Sir James   13–14, 24n.20
Land Development Control Plan   198
land holding   31–3, 44, 50n67, 63–4, 72, 81
Lane, Colonel William   84, 140
Las Cases, Count de   152–3
Lefroy, Lieutenant J. H.   122
Leguat, François   26
Liberated African Depots   168, 170–2, Figs 25–6
lime burning   59
Linschoten, Jan Huygen van   8, 10, 21, 27, 67n.11
Liverpool, Lord   151, 164
lookouts   72n22
Lopes, Odoardo Duarte   9n.5
Lopez, Fernão   9, 18, 34
Lowe, Sir Hudson   40, 61, 126, 156–60, 162–3, 164n.35, 208, Fig. 22

Madagascar   36
'Madagascar ships'   36
Madeira   191
magnetic surveys   119–24, 146, 202
Maitland, Captain Frederick   150
Malay presence on St Helena   77n40
Marden, Major S. H.   177
Marshall, Sarah   37
Marryat, Captain Frederick   160, 162, Fig. 23, Plate 24a
Mascarene Islands   5
Maskelyne, Revd Nevil   113–15
Mason, Charles   113
Mason, Polly   49n.62
Massingham, Sir John   208
Mauritius   30, 143, 193
Mayer, Erich   Plate 25
Melling, William   37n.34
Mellis, Captain G. W.   147
Mellis, John Charles   147–8
Mercury, Transit of   108, 111, 113
meteorological observation   124
Middlemore, Major-General George   53, 80, 167n.8, 208
Military Institute   115
Millennium Gumwood Forest Project   203, Plate 30

mineral resources (submarine)   204
model of St Helena's topography and
    defences   79–80, 146n.97
Molesworth, Richard   50
Montholon, Count and Countess   152, 155
Moore, Sir Jonas   109
Morris, Dr D.   51
Mosley, Alfred   56
Munden, Captain Richard   35n.27, 70, 72, 207
Mundy, Peter   17, 19n.5, 20–2, 25–7, 66–7
Murphy, Sir Dermod   208
mutinies, revolts   36n.28, 72, 180n.30

Namibia   200
Napoleon Bonaparte   33, 42, 51n73, 61, 79, 115, 126, 150–66, Fig. 23, Plates 21–5
    arrival on St Helena   152
    household   152
    accommodation
        in Jamestown   152–3
        at The Briars   153, Plate 21
        at Longwood House   154, Plate 23
    relations with Sir Hudson Lowe   157–9
    final illness and death   159–60 Fig. 23
    funeral and burial   162–4, Plate 24
    exhumation and transfer to
        France   165
Navarette, Fernandez   42n44
Ndabuku   172
New Zealand   54
Newton, Sir Isaac   120
Nicholls, Alan   189
Nissen, Colonel P. N.   185
Nova, João da   9–10

Oates, Sir Thomas   208
Ólafsson, Jon   19n.5, 25
Oldenburg, Henry   107
O'Meara, Barry   150n.2, 152, 158n.20
Oxendon, Sir George   36n.29

Pacific Ocean   5–6, 12n15, 126, 128, 131, 139, 145
Palmer, Captain Edmund   1
paper roofing   33
Patey, Vice-Admiral Charles   54, 208
Patton, Colonel Robert   26n.29, 38n.38, 43, 62, 74, 83–4, 133, 137, 140, 207

Peel, Colonel Sir Robert   61, 208
Phillips/Honan, Lisa   208
Phillips, Nigel   208
Pickard-Cambridge, Revd O.   147
pigments   141
Pilling, Sir Guy   208
Pine Coffin, John   208
plantations and nurseries   50, 51, 53, 61
plants   4, 14, 18–21, 31–2, 48, 50, 53, 61n.102, 126–8, 131–3, 136–7, 139–47, 159n.25, 203
    American aloes   64
    aniseed   19
    anti-scorbutic herbs   13, 20
    apples   19
    artichokes   129
    asparagus   129
    asplenium   136
    barley   45, 49n66
    basil   19
    beans   21, 40n.39, 47, 129
    Bermuda cedars   63
    blackberries   62, 145, 170, 203
    brambles   136
    breadfruit   51, 128, 130–1
    cabbage   48, 129
    Cabbage trees (*Senecio redivius, Senecio leucandron, Melanodendron integrifolium*)   4, 127, 130, 136
    cactus (*Opuntia* sp.)   50, 141n.81
    camomile (*Cotula anthemoides*)   19n.5
    campanula   136
    carrots   129
    cassava   21, 31, 47
    cereals, grains   26, 31, 50
    cherries   51
    chickpeas   21n.12
    cinchona (*C. succirubra* and *C. officinalis*)   53
    clover   12n.60
    coconuts   51
    coffee   51, 64
    cotton   64
    cress   129
    cucumbers   129
    currants   51
    dates   19
    ebony (*Trochetiopsis ebenus*)   4, 59–60, 128, 149

Index

fennel   19
ferns   17, 136, Maiden-hair fern (*Adiantum caudatum*)   113, Figs 19, 51
flax (*Phormium tenax*)   54, 64, 203, Plate 11
gladioli   137
gooseberries   51
gorse, furze (*Ulex europaeus*)   61, 136, 145, 170
gourds   9
grapevines   52, 140
grasses (*Eragrostis episcopulus*)   4, 17, 129
   cow grass   62
   lucerne   62, 129
   sanfoin   129
   vernal grass   62
   wire grass   62
guavas   51
gumwood   6, 57n.88, 61, 127, 130, 137, 203, Plate 30
Indian corn   45, 129
indigo   36
ixias   137
larch   139
lemons   19-21, 57n.88, 61
lichens   141
lobelia   136
Lombardy poplars   61
lonchitis   136
mallows   19n.5
Marchantia   136
melons   12
Mexican pine   63
mint   19n.5
mountain rice   36, 51
mulberries   51
mustard seed   19, 129
Norfolk Island pine   63
oaks   61
   live oaks   130
oats   45, 49n.66
olives   51
oranges   19, 21
palms   4n.7, 9
parsley   19
peas   21, 47, 129
'peeah'   132

'physic nuts'   51
pineapples   61, 140
pineaster   61
pines   61
plantains   21, 31
pomegranates   9
potatoes   21, 45, 47, 52, 129
pulses   21n.12
radishes   19, 129
redwoods (*Trochetiopsis erythroxylon*)   4, 17, 59–60, 136
sago   51
Scotch firs   61
snail trefoil   129n.60
sorrel   19
spruce   61
sugar cane   36, 53
thatching rush (*Fimbristylis textilis*)   33n.15
tobacco   19n.5, 36
tree-ferns (*Dickinsonia arborescens*)   4, 136, Plate 20b
wheat   45, 49n.66, 64
willows   145
yams   17, 21, 36, 44–5, 47, 60n.100, 127
ploughs   49–50
Poirier, Captain Stephen   52, 73, 207
Poivre, Pierre   143
Porteous, Henry   132, 153
Portlock, Lieutenant   131
Portuguese presence on St Helena   9–11, 14–15, 18–21, 25n.25, 34, 65, 68, Fig. 4
postage stamps   64, 148, Fig. 19i
'pounding days'   58
Powell, George   207
Prince Regent (later King George IV)   151, 188
Prior, Lieutenant James   47, 76
prison   62
Pritchard, Captain Henry   158n.22
Pyke, Captain Isaac   30, 34, 37, 44, 51, 60, 62, 207
Pyrard, François   14, 19

Radio St Helena   192
Ramage, Sergeant   117
Raughan, Mr   132
Rennefort, Sieur de   24, 32n.10, 68
Rennel, Major James   74, 130, Plate 3

225

Index

roads 48, 76
Roberts, Captain John 32–3, 36, 44, 56–7, 58n.90, 60, 73–4, 207
Roberts, Mr 139
Robson, Major 132
Rogers, Francis 20n.10, 27, 67
Roggeveen, Jacob 12n.15
rollers 77, 124–6, 194, Plate 16
rooftiles 32
Rose, Mr 139
Ross, Captain Charles 152
Ross, James Clark 122, 146
Ross, Major-General Sir Patrick 208
Roxburgh, William 33n.15, 139, 143–5, 146n.96, 147, Fig. 20b
Royal African Company 36
Royal Astronomical Society 117
Royal Fleet Auxilliary 82
Royal Marines 81, 162
Royal Navy 81–3, 152, 154, 162, 167, 183, 194
    West Africa (Preventative) Squadron 167–72
Royal Society 107, 112–14, 115n.22, 120, 121–2, 124, 126, 141, 146
Rushbrook, Dr Philip 209

St Helena, geology, mineralogy and palaeontology 3, 141
    as Utopia/Eden 6, 10, 31, 52n.78, 127–8, 204–5
    deforestation and degradation 52n.78, 59, 128, 136, 146
    as a node in plant exchange 128–33, 139, 205
    climate, droughts and floods 3, 33–4, 60–1, 76
    fertility 48, 127
    as a prison 150–89
    reversion to the Crown 41, 53, 147, 166n.1, 194
    as a British Dependency 190
    as a British Overseas Territory 190
    Alarm House 83, 114
    Asses Ears 149, 200
    Banks's Battery 71, 73, 76
    Bilberry Field Gut 57n.88
    Breakneck Valley 44
    The Briars 51, 153, 158, 165, 191–2

Broad Bottom 180
Brooke Hill Farm Fig. 7
Bryant's Beacon 193
Buttermilk Point 76
Chapel Valley (later James Valley) 10, 17, 19–20, 25, 65, 67–8, 71, Fig. 3
Deadwood Plain 62, 155, 157–8, 174–5, 180
Diana's Peak 3, 53, 62, 136, 139, Plate 5
Diana's Ridge Plate 19
Dry Gut 200
Egg Island 22, 141
Farm Lodge 153n.8
Flagstaff Bay 76
Fort James 68, 71, 73–4, Figure 11
Francis Plain 181
Francis Plain House 172
Friars 44
Geranium Valley 162
Great Wood 4, 26n.30, 56, Plate 30
Half Tree Hollow 175
Halley's Mount 109
High Knoll 84, 175, 177
Hutt's Gate 154
James Valley (formerly Chapel Valley) 67, 74, 114, 132, 158, Figs 3, 13
Jamestown 32, 47, 50, 54, 56, 63–4, 70, 72, 74, 76–7, 83, 116, 121, 124, 137, 152–3, 158, 167, 172, 181, 183, Fig. 28
Keigwin's Point 70
Kent Cottage 175
Knoll Hill 74–5, Plate 13
Knollcombes 181, Fig. 31
Ladder Hill 37, 49n.62, 58, 74, 77, 81, 115–19, 147, Fig. 17, Plate 14
Lemon Valley 20, 68, 71, 74, 76–7, 121, 124n.51, 168
Little Stone Top 136
Longwood Observatory 122–4, Fig. 19
Longwood Plain 45, 50, 79n.43, 124, 153–4, 155, 157, 160, Fig. 9, Plate 23
Longwood House 33, 153–5, 158, 162, 165, Plate 23
New Longwood House 154
Manatee Bay 28
Munden's Hill 82n.55
Munden's House 187
Munden's Point 73–5, 187

Index

New Ground 44
New Longwood 45
Oakbank 153n.8
Oaklands 153n.8
Old Woman's Valley 68
Parsley Bed Ridge 71
Patton's Battery 74
Plantation House 50, 60, 61, 132, 137, 139, Plate 10
Plantation Valley 57n.88
Prosperous Bay 35n.27, 67, 70, 77, 79n41, 83, 199–200
Rose Cottage 177
Rosemary Hall 126, 153n.8, 172
Rupert's Bay 67, 71, 74, 76, 191–2, 198, 200, Fig. 10
Rupert's Valley 62, 168, 170
Sandy Bay 59n.95, 76
Sandy Ridge 136
St Helen's Chapel 6, 10, 12, 14, 68
St Paul's 172, 192
Sandy Bay 29, 60, 67
Spring's Platform 71
Thompson's Bay 77
St Helena Airways 199
St Helena Climate Change and Drought Warning Network 203n.19
St Helena Government Broadcasting Station 192
St Helena Industries Ltd 56
St Helena Leisure Corporation 199
St Helena Relief Committee 148
St Helena Research Institute 203
St Helena Shipping Company 195, 197
St Helena Whale Fishery Company 55, Plate 12
St Jago (Santiago) 21, 31, 35
St Paul de Noronha 125
St Vincent 133, 191
Sabine, Major Edward 122, 124
Saint FM 192
salt rations 20n10, 33–4, 38, 53, 60
salutes, firing of 73–4
satellite communication 192
Sayyid Khalid bin Bargash Al-Busa'ldi 182–5, Fig. 32
scurvy 13, 20n10
Seale, Robert Francis 80, 146n.97
Segar, John 13, 24

Senegal 129
Seychelles 185
shellfish and crustacea 29
shingles 33
ships, sailing 9, 13, 15–16, 30, 32n.9, 47, 53, 65, 67, 70, 71n.17, 76n.37, 77, 83–4, 108, 131, 139, 141, 143–4, 167, 191, 193
    Baltimore clippers 168
    steamships 81, 193–4
    Alfred 139
    Anglia 191, Fig. 33
    Assistance 70
    Assistant 130–1
    Beagle 121n.39, 144–5
    Beaver 160
    Bellerophon 150–2
    Benjamin 52
    Brecon 195, Fig. 34
    Canada 182
    Castle Huntly 144
    Cawdor Castle 185
    Ceres 139
    Chanticleer 120–1
    Cordelia 125
    David Scott 139
    Decobrador 125
    Darkdale 82
    Defence 20
    Drake 43n.47
    Erebus 146
    Earl of Ashburnham 130
    Endeavour 49, 126
    Espoir 121
    Europe 139
    General Goddard 76
    Glatton 139n.77
    Golconda 182
    Golden Fleece 112
    Great Britain 193–4
    Helena 198
    Heron 162
    Humphrey and Elizabeth 68, 70
    Ledbury 195, Fig. 34
    Loch Insh 187
    London 21, 31
    Marmaduke 30
    Milwaukee 174
    Nisus 76
    Newcastle 152

Index

Northland Prince   195
Northumberland   125
Ocean   50n68
Paramore   57, 112n.11, 120
Pearl   19n.5
Pensacola   124–5
Phaeton   156
Prince Henry   114
Providence   130, 133
Repulse   51
Resolution   49, 128
Rosario   162
Royal Merchant   13
St Helena (pre-1830)   193
St Helena (1978–1990)   194, Fig. 34, Plate 27
St Helena (1990–2018)   194–7, Plate 28
Sceptre   76
Terror   122, 146
Three Brothers   139n.78
Truro   35
U-68   82
Unity   108
Walmer Castle   139
Waterwitch   167
Witte Leeuw   15, Plate 7
Sierra Leone   167, 172, 193
Siglé, H. T.   177, Fig. 29
signals, signalling posts and systems   73n.28, 83–5, 117, 119, 158, Plates 15, 22
silk industry   51
Sirius, Parallax of   113–14
Skottowe, John   129, 207
Slave Emancipation Act   (1834) 34
Slave Trade, Act for the Abolition of (1807)   167
Slavery Abolition Act (1833)   167
slavery, enslaved persons   17, 34–45, 50, 53, 72, 167–72
   from Africa   34–6, 38n.39, 167, 170
   from Burma   36n.30
   from France   42n.44
   from India   36n.30, 42–3
   from Java   34–5
   from Madagascar   36
   from the Maldives   43n.47
   from Portugal   42n.44
Smallman, David   208

Smith, Christopher   131–2
Smith, Captain John   207
Solander, Daniel   126
South Africa   5n.12, 91–2, 174, 182, 191–4, 194, 197–9, 202
   see also Cape of Good Hope
Spanish presence on St Helena   65
Specx, Jacques   15, Plate 8
Stack, Lieutenant F. R.   Plate 16
Sténuit, Robert   15
Sterndale, Robert   175, 208
Stimson, Robert   208
stone, as building material   32
storm damage   76–7, 125
straw work   64
Stringer, Captain Robert   31
Stürmer, Baron von   126
submarine telegraph   85, 190–2
Suez Canal   194
Sure South Atlantic Ltd   193
Sustainable Cloud forest Restoration Project   203
Swanley, Captain   35

Tahiti   130–2
Tahitian presence on St Helena   132
tanning industry   59
Tavernier, Jean-Baptiste   19, 67
telegraph stations   84–5, 158
Tenerife   195
'terra Puzzolana'   32–3
thatch   33, 45
tidal surveys   115n22, 117
timber, as building material   32
Time Ball Office   84
time keeping   113–15, 117–21, Fig. 18
trade winds   3, 61, 111n.10, 112, 124, 191
Trelawney, Colonel Hamelin   208
Tristan da Cunha   3, 112n.11, 190n.1, 197, 204
Tshingana   172

Union-Castle Line/Mail Steamship Company   193–4

Index

United States Scientific Expedition to West Africa   124
Union Steamship Company   194

Valentia, George Annesley, Viscount   47, 49n.62, 132n.66
Venus, Transit of   112–14
Vice-Admiralty Court   167
Viljoen, General Ben   177
vineyards   52

Waddington, Robert   114, 115n.22
Walker, Alexander   40, 45, 50, 81, 115, 208, Fig. 17
Walvis Bay   200, 202
water resources   9n.6, 17, 19, 56–7, 159, 203, Fig. 2

wax   64
Webster, William   28, 66, 117
Welle, Philippe   126
Wellesley, Sir Arthur (later Duke of Wellington)   153n.7, 157n.16
West Indies   36, 45, 53, 59n.96, 130–2, 172, 191, 205
whale fishery   54–5
whale oil industry   Fig. 10
wheelbarrows   49
Wiles, James   131–2
Wilks, Colonel Mark   50, 61, 153, 155, 158, 208, Plate 22
Windhoek   202
windmills   33, 45, 177, Figs 9, 29

Zanzibar   182–3

Printed and bound by CPI Group (UK) Ltd, Croydon, CR0 4YY
07/07/2024
14524623-0002